Planets and Moons

Planets and Moons

William J. Kaufmann, III

Department of Physics
San Diego State University

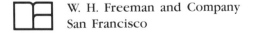
W. H. Freeman and Company
San Francisco

Sponsoring Editor: Arthur C. Bartlett; *Project Editor:* Pearl C. Vapnek; *Copyeditor:* Karen Judd; *Designer:* Marjorie Spiegelman; *Production Coordinator:* Linda Jupiter; *Illustration Coordinator:* Batyah Janowski; *Artist:* Dale Johnson; *Compositor:* Graphic Typesetting Service; *Printer and Binder:* The Maple-Vail Book Manufacturing Group.

Library of Congress Cataloging in Publication Data

Kaufmann, William J
    Planets and moons.

    Bibliography: p.
    Includes index.
    1. Solar system. I. Title.
QB501.2.K38   523.2   78-21156
ISBN 0-7167-1041-2
ISBN 0-7167-1040-4 pbk.

Printed in the United States of America

9 8 7 6 5 4 3

Cover photograph: NASA.

to Lynne E. with love

Contents

Preface

To be the first to see what no one has ever seen; to go where no person has ever gone; to experience, to think, and to feel that which is not yet in the repertoire of human consciousness. This is the inspiration of the scientists who reach for the planets and beyond to the stars. It is a true frontier populated with pioneers. But these explorers are totally unlike any of their predecessors. Years of patient, careful planning are followed by meticulous construction of bizarre vehicles in sterile, dust-free rooms. Then comes the long wait, the coasting across the black void of interplanetary space. And it all pays off—the labors, the expenses, the anxieties, the frustrations—as mechanical eyes with robot hands come to life and a transistorized voice begins to speak. Few experiences are more exciting than being present as we receive our first view of the craters of Mercury or dig our first trench on Martian soil.

This is what it means to live in the twentieth century. We have walked on the moon and seen beneath the clouds of Venus. And we are all part of this venture. For thousands of years, people have gazed up at our satellite and seen those familiar man-in-the-moon features that perpetually face our planet. Yet *we* were alive when humanity first saw the hidden side of the moon. For a one-time, per capita expenditure of *less* than the cost of a single pack of cigarettes, *we* financed a voyage past the rings of Saturn. We live in a time of adventure and exploration on the grandest scale. What we find shall surely prove as far-reaching as the discoveries of Columbus and Magellan.

This is our legacy. And just as the revelation of new continents pro-
foundly affected the course of Renaissance Europe, our interplanetary
explorations will certainly play a significant role in shaping the future
of humanity for centuries to come.

October 1978   William J. Kaufmann, III

Planets and Moons

1

In the Beginning

It must have been cold — incredibly cold — 5 billion years ago. Right here, where there are now trees and streets and people — right here in our own tiny corner of the galaxy. But long ago — so very long ago — before the birth of the sun and the creation of the planets. Stretching for billions upon billions of miles in all directions — the sparse interstellar medium — a frigid, nearly perfect vacuum in the blackness between the ancient stars.

It must have been cooler than 50 degrees above absolute zero. For comparison, "room temperature" is nearly 300 degrees above absolute zero, and the oxygen in the air we breathe liquefies at 90 degrees above absolute zero. But the primordial interstellar gas was in no danger of freezing or liquefying; the atoms were so widely spaced that they had too little opportunity to collide and stick to each other.

It must have been a nearly perfect vacuum: only a dozen atoms per cubic centimeter (which is the same as 200 atoms per cubic inch). For comparison, the air we breathe contains roughly 30 million trillion atoms per cubic centimeter. A space traveler would hardly have been aware that he or she was in the midst of a huge, primordial cloud of gas and dust from which our solar system would eventually be born.

Hydrogen was by far the most abundant substance. Nearly three-quarters of the interstellar cloud, by weight, consisted of hydrogen. And almost one-quarter of the cloud, by weight, was helium. In terms of numbers, this means that there was one helium atom for every dozen hydrogen atoms.

This preponderance of hydrogen and helium in interstellar space completely overshadowed the abundances of all the heavier elements. Since over 95 percent of the mass of the interstellar cloud consisted of hydrogen and helium, only a few percent was left over for all the heavier elements combined. Some of these heavier elements existed in the form of very tiny dust particles, typically a thousandth of a millimeter in size. But these substances were so rare that the dust grains were few and far between. A space traveler would have found only a hundred of these microscopic dust particles in an entire cubic kilometer (which is roughly the same as 400 per cubic mile) inside the interstellar cloud.

Figure 1-1   The Orion Nebula
*Stars and planets are created in huge interstellar clouds of gas
and dust. At the time of star-creation, the cloud is very cold and
dark. But after star-birth, radiation from the young stars can
cause the cloud to glow with unprecedented beauty. Newborn stars
are embedded in this spectacular nebula in the constellation of
Orion. (Lick Observatory.)*

These widely spaced dust grains consisted mostly of silicon, magnesium, aluminum, and iron—exactly the same substances from which ordinary rocks are made. But, in addition, other familiar elements such as oxygen, carbon, and nitrogen occasionally existed in the form of organic molecules. Dozens of different organic molecules are found in interstellar space. This means that the chemical building blocks of life were present long before our sun and the planets began to form.

There are two theories about how the solar system began. The primordial interstellar cloud could not start forming the solar system by itself; the cloud was simply too diffuse. Something must have happened to compress the cloud.

We live in a spiral galaxy whose overall appearance is similar to the galaxy shown in Figure 1-2. Some astronomers believe that a spiral arm of our galaxy passed through our region of space some 5 billion years ago. This would have caused a slight compression of the interstellar cloud, and star-creation could have begun. Indeed, we find many young stars and glowing gas clouds outlining the spiral arms of distant galaxies.

Other astronomers believe that a nearby massive star became a supernova. During the final few hours of its existence, this ancient and unknown star was torn apart in nature's most cataclysmic detonation. The resulting shock wave would have been sufficient to compress our interstellar cloud, and star-creation could have begun. The remains of a star that became a supernova only 20,000 years ago is shown in Figure 1-3. Similar nebulosity of the supernova that started the sun's formation has long since disappeared. Nevertheless, scientists analyzing meteorites have recently discovered unusual abundances of certain elements that would have easily been produced by a nearby supernova explosion.

Prior to compression, our primordial interstellar cloud had been in equilibrium. The force of gravity, which tried to make the cloud contract, was exactly balanced by the gas pressure inside the cloud. But after compression (either by passage through a spiral arm or from a supernova explosion), the microscopic dust grains in the cloud were squeezed closer together than before. After compression, there might have been as many as 10,000 dust grains per cubic

Figure 1-2   A Spiral Galaxy
*Our galaxy—if we could see it from a great distance—would
look like this galaxy in the constellation of Ursa Major. Our
galaxy contains over 100 billion stars and measures 100,000
light years in diameter. We are located two-thirds of the way from
the center, between two spiral arms. (Kitt Peak National
Observatory.)*

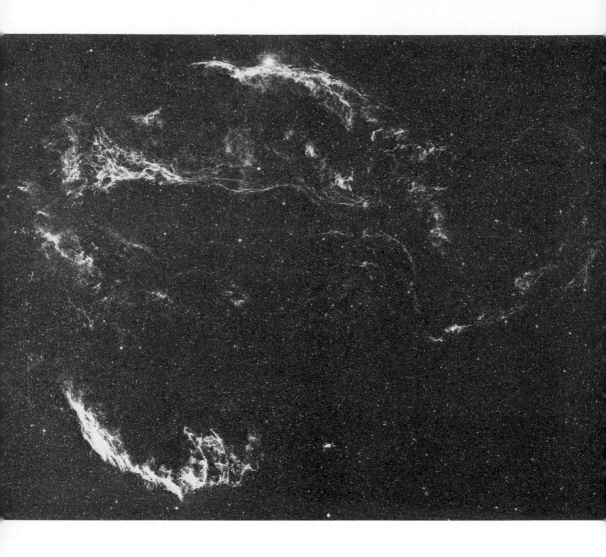

Figure 1-3   A Supernova Remnant
Massive stars end their life cycles with supernova explosions.
These wispy, glowing clouds in the constellation of Cygnus are
from a star that blew itself apart 20,000 years ago. The nearly
spherical nebula is 120 light years in diameter. Shock waves from
supernova explosions can compress cool interstellar clouds and
start star-creation elsewhere in the galaxy. (Hale Observatories.)

8

kilometer—a hundredfold increase in the density of dust. This increase had the immediate effect of shielding the inner portions of the interstellar cloud. Light from nearby stars could no longer shine through the cloud.

The obscuring effect of interstellar dust grains played an important role in the genesis of our solar system. Since starlight could no longer penetrate and warm our cloud, the temperature of the gas inside the cloud began to plunge toward absolute zero. Gas pressure and gas temperature always go hand in hand. Consequently, as the temperature declined, the pressure that the gas could exert also decreased. The outward gas pressure of the cloud was no longer able to resist the inward push of gravity. Gravity won and the cloud began to contract.

Astronomers often find cool, dark, contracting clouds of interstellar gas and dust that are at the initial stages of star-creation. As shown in Figure 1-4, these so-called globules are most easily seen when silhouetted against a bright nebulosity. A typical globule is a few light years in size and contains enough matter to make dozens of solar systems.

As our globule contracted under the influence of gravity, eddies and slowly rotating whirlpools began to develop from random turbulence within the cloud. These eddies caused the cloud to break up into smaller peices. One of these slowly rotating cloud fragments was destined to become our solar system.

As our cloud fragment continued to contract, its rate of rotation began to speed up. This rotation caused our cloud fragment to become distinctly disk-shaped. This was the *primordial solar nebula*. It measured 10 billion kilometers across (roughly the same size as Neptune's orbit), was nearly 200 million kilometers thick (roughly the same as the distance from the earth to the sun), and contained twice as much matter as is presently found in the solar system.

Gravity continued to dominate the early evolution of the primordial solar nebula as more and more matter contracted toward the center of the disk. This infalling gas caused the central regions of the solar nebula to become significantly hotter than the outer regions. The interstellar dust grains in the inner regions were soon

9

Figure 1-4   Globules in the Lagoon Nebula
*Numerous tiny globules can be seen silhouetted against this
bright nebulosity in the constellation of Sagittarius. These dark,
cool clouds of gas and dust are contracting under the influence
of gravity. In only a few hundred million years, many of them
might become newborn stars. (Kitt Peak National Observatory.)*

completely vaporized. The enormous temperature difference between the center and edge of the solar nebula would eventually have a profound effect on the solar system. The inner planets were destined to be very different from the outer planets.

Fifty million years after that fateful compression of the interstellar cloud, formation of the solar nebula was complete. Material continued to plunge to the center of the nebula, and the protosun was formed. During this time, the sun's primeval magnetic field kept the protosun in contact with gases throughout the rest of the solar nebula. Without this contact the sun would have ended up rotating at a furious rate — just as an ice skater doing a pirouette can speed up dramatically by pulling in his or her arms. But our sun is rotating fairly slowly, only once every four weeks. The dragging of the protosun's magnetic field through the gases in the solar nebula must have had a severe breaking effect. Rotation was therefore more evenly distributed throughout the solar nebula rather than being concentrated in a rapidly spinning sun. This stage, during which rotation was transferred from the inner to the outer portions of the solar nebula, lasted for only a few thousand years. When the transfer was complete, the planets were ready to be born.

Material in the primordial solar nebula could be divided into three broad categories based on melting or boiling points. First of all, there were substances normally associated with *rocks*. These include silicates, oxides of metals, silicon, magnesium, aluminum, and iron in various chemical combinations. All of these substances have very high melting or boiling points, typically at temperatures of thousands of degrees.

Secondly, there were substances normally associated with *liquids and ices*. These mostly involve chemical combinations of carbon, nitrogen, hydrogen, and oxygen. Perhaps the most familiar of these substances include water, carbon dioxide, methane, and ammonia. All have melting or boiling points in the range of 100 to 300 degrees above absolute zero.

And finally, there were substances that are almost always *gases*. These include hydrogen, helium, neon, and argon in their pure form. Except at incredibly cold temperatures near absolute zero, these substances are always gases.

These temperature considerations played a crucial role in determining the nature of the planets that formed at various distances from the sun. During the creation of the protosun, as vast quantities of matter plunged to the center of the primordial solar nebula, temperatures surrounding the center of the nebula were quite high. Temperatures of several thousand degrees were common, and all substances were completely vaporized. But in the outer portions of the nebula, the temperature was never much higher than 100 degrees above absolute zero. Rocky dust grains in these outer regions were probably coated with water ice, dry ice, and frozen methane and ammonia. These remote, ice-coated grains were largely undisturbed by the gravitational contraction of the protosun.

Following the formation of the protosun, temperatures in the inner regions of the solar nebula began to decline. As the gases cooled, substances could begin to condense out of the solar nebula. Rocky substances were naturally the first to solidify from the vaporized state. But near the protosun, temperatures always remained sufficiently high so that only the rocky substances could solidify. Dust grains nearest the protosun therefore consisted mostly of iron, silicates, and oxides of metals.

A little farther away from the protosun, temperatures were still lower. A thin layer of ices could coat the rocky dust grains located at these moderate distances. And at great distances from the protosun, the dust grains were coated with ample layers of ices. All of these dust grains—both near and far—were still embedded in a huge cloud of hydrogen and helium, the two abundant gases that together constituted more than 95 percent of the solar nebula. But at this stage there were, for the first time, significant differences in the composition of the grains at various distances from the protosun.

The dust grains in the solar nebula were probably rather fluffy. Like large snowflakes, they easily stuck to each other after colliding. Over the years, repeated collisions produced clumps typically measuring a few millimeters or centimeters in diameter. Gradually, under the influence of gravity, these clumps settled toward the midplane of the solar nebula.

This settling process lasted for a few hundred thousand years. At the end of this stage, most of the solid material in the solar system was distributed in a huge sheet with the protosun at the center. But a vast sheet of tiny clumps is unstable under the influence of gravity. More clumps are attracted to locations in the sheet where, by chance, there already is a slight excess of clumps. And clumps wander away from those locations in the sheet where, by chance, there initially was a lower than average number of clumps. In this way the clumps gradually amalgamated into kilometer-sized, asteroidlike objects called *planetesimals*.

It is important to realize that the chemical composition of the planetesimals varied greatly across the solar nebula. Near the protosun, the planetesimals were made almost entirely of rocky substances. This is because the initial grains (and later clumps) contained only those substances that could remain solid in the warm, inner regions of the primordial solar system. Farther from the sun, where the temperature was lower, water ice was included with the rocky compounds. And in the cool, remote regions, planetesimals also contained frozen methane and ammonia.

Gradually over the next few million years, the mutual gravitational attraction between planetesimals caused them to coalesce and consolidate into much larger objects called *protoplanets*. Four protoplanets formed in the inner regions of the primordial solar system. And four more protoplanets formed at much greater distances from the protosun. As we shall see in Chapter 8, there is reason to believe that Pluto—now recognized as the smallest planet in the solar system—was originally a satellite of Neptune.

The four inner protoplanets were destined to become Mercury, Venus, Earth, and Mars. The decay of radioactive isotopes inside the protoplanets soon heated and eventually melted their interiors. Once again the force of gravity came into play as the heavier substances (mostly iron) sank to the centers of the molten protoplanets. This pushed the lighter substances up toward the surfaces. In this way the planets became "chemically differentiated" with dense iron cores surrounded by less dense layers of rock.

During these ancient times, when the four inner planets were in essentially molten states, gases readily escaped from the liquid rocks. Mercury, the second smallest planet in the solar system, was totally incapable of holding onto any of these gases. All the gases

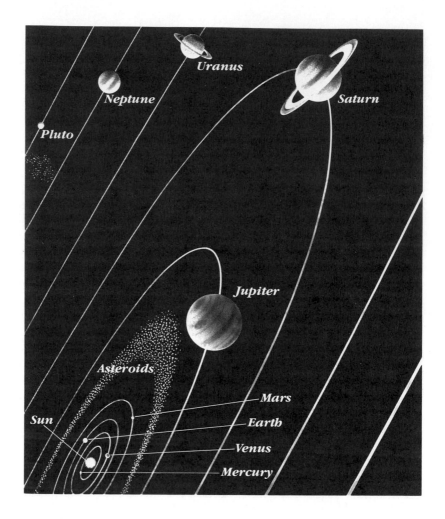

Figure 1-5   The Solar System

This scale drawing of the solar system shows the spacing of the orbits of the planets. Notice that the four inner planets are crowded close to the sun. By contrast, the orbits of the outer planets are spread out over much larger distances.

soon boiled off into space under the combined influence of the searing heat from the young sun and Mercury's low surface gravity.

Mars, the third smallest planet in the solar system, also has a very lower surface gravity. Consequently, Mars also lost much of its primordial atmosphere. Only a very thin layer of carbon dioxide gas remains.

Only Venus and Earth, the largest of the inner planets, possessed surface gravities strong enough to retain atmospheres. But these atmospheres are extremely scanty — a layer of gases huddled next to the ground. Most of the atmospheric gases surrounding Venus and Earth are at elevations lower than 10 kilometers above the planets' surfaces. This stands in sharp contrast to the outer planets, which have atmospheres tens of thousands of kilometers thick. The primary reason for this disparity is directly related to the chemical compositions of the grains from which the planets formed. Grains in the warm, inner portions of the solar nebula had little or no coatings of icy substances. Consequently, the inner four planets — like the grains from which they are formed — are almost entirely composed of rocky substances. At first glance, the tiny smattering of gases and liquids around some of the inner planets is hardly worth mentioning.

Chemical differences in the primordial grains also played an important role in determining the interior structure of the four inner planets. All four planets have iron cores surrounded by mantles of less dense rock. But compared with the outer planets, Mercury clearly has the largest iron core. Mercury's iron core extends three-quarters of the way from the center of the planet to its surface. Mercury's iron core accounts for 80 percent of the planet's mass. By contrast, the iron cores inside Venus and Earth extend only halfway from their centers to their surfaces. And the iron core inside Mars is even smaller.

Iron, nickel, and oxides of other metals were the first substances to condense from the hot inner regions of the primordial solar nebula. These substances have the highest condensation temperatures. Silicates and other rock-forming minerals condense at slightly cooler temperatures. Consequently, grains that condensed near the protosun contained a higher abundance of iron compared with the more distant grains. Thus, the planet that formed nearest the sun contains the highest percentage of iron. More distant planets contain lower percentages.

The creation of the giant outer planets must have begun at nearly the same time and in the same way as the inner planets. But the planetesimals in the distant, cold regions of the solar nebula contained substantial quantities of ices. Planets that formed in these remote regions would be destined to have thick atmospheres of methane, ammonia, and other gases.

In the case of Jupiter and Saturn, the buildup and amalgamation of planetesimals was so effective that the strong gravitational fields of these huge protoplanets could easily attract hydrogen and helium. In addition to retaining all their gases, proto-Jupiter and proto-Saturn swept up vast quantities of hydrogen and helium as they orbited the young sun. Indeed, the creation of Jupiter and Saturn must have mimicked the formation of the solar system itself. Like miniature solar systems, each of these giant planets is surrounded by a substantial retinue of satellites.

The buildup and consolidation of planetesimals was not quite as pronounced in the case of Uranus and Neptune. Their protoplanets, although very large compared with the inner planets, never grew to the enormous size of Jupiter or Saturn. Uranus and Neptune could only accrete a small amount of lightweight hydrogen and helium from the solar nebula. The thick atmospheres of Uranus and Neptune therefore contain less hydrogen and helium than Jupiter and Saturn do. But like their giant neighbors, Uranus and Neptune are surrounded by satellites. Pluto, now a planet, may have originally been a satellite of Neptune.

While the planets were forming from materials in the solar nebula, the protosun continued to evolve. The inward pressure of trillions upon trillions of tons of gas caused the center of the contracting protosun to get hotter and hotter. Finally, 4½ billion years ago, temperatures at the sun's center became so high that thermonuclear fires could be ignited. This ignition of thermonuclear processes—the fusing of hydrogen into helium at temperatures of millions of degrees—signals the birth of a star. Our sun was born.

Sun

Moon

Venus **Mars**

Mercury **Earth**

Jupiter

Saturn

Uranus

Neptune

Pluto

Figure 1-6   The Planets
*This scale drawing shows all the planets and a portion of the sun.
The four inner planets are very small and have hard, rocky surfaces.
The next four planets are gigantic and consist almost entirely of
gases and liquids.*

Astronomers often find young and newborn stars in the sky. A whole cluster of very young stars is shown in Figure 1-7. Many of these stars are now in the process of igniting the thermonuclear fires at their cores.

From carefully observing very young stars, astronomers now realize that stars often eject substantial quantities of matter at the end of the birthprocess. As a newborn star adjusts to the onset of thermonuclear reactions in tis core, a large amount of gas is blasted away from the star's surface. This ejection of matter is called a *T Tauri wind*.

It is reasonable to assume that all stars are surrounded by stellar winds. These "winds" actually consist of a continuous emission of particles—mostly protons and electrons—from the star's surface. Our sun today continuously sheds particles, producing the *solar wind*. The discovery of the solar wind in the early 1960s by the first interplanetary spacecrafts is among the most important revelations of the entire space program. At the earth's orbit, the average speed of the solar wind is 400 kilometers per second (nearly a million miles per hour). The average density of the solar wind in our vicinity is a meager ten particles per cubic centimeter. But the solar wind is sometimes rather "gusty." Densities of nearly 100 particles per cubic centimeter and speeds approaching 1,000 kilometers per second have been recorded by spacecrafts journeying toward the planets.

The stellar winds that surround middle-aged stars and the sun are gentle breezes compared with the T Tauri wind. The T Tauri wind is a veritable hurricane, a powerful gale that exerts a substantial pressure on everything in its path.

The T Tauri wind that accompanied the birth of our sun blasted all the remaining excess hydrogen and helium far out into interstellar space. The initial primordial solar nebula contained enough matter (mostly hydrogen and helium) to make *two* suns. But during the million years that the T Tauri wind raged across the young solar system, nearly half the primordial gases were returned to the depths of space.

The T Tauri wind cleaned up the solar system. The gale was so strong that the inner planets were stripped of most of their primordial atmospheres. Only solid particles—planets, satellites, asteroids,

Figure 1-7 A Cluster of Newborn Stars
This cluster of stars and gas (called NGC 2264) is a stellar nursery in the constellation of Monoceros. Many of the stars in this cluster are only now igniting thermonuclear reactions at their centers. Many of these stars are passing through their "T Tauri" stage. These stars are producing powerful stellar winds that are blowing away the nebulosity from their surroundings. (Hale Observatories.)

19

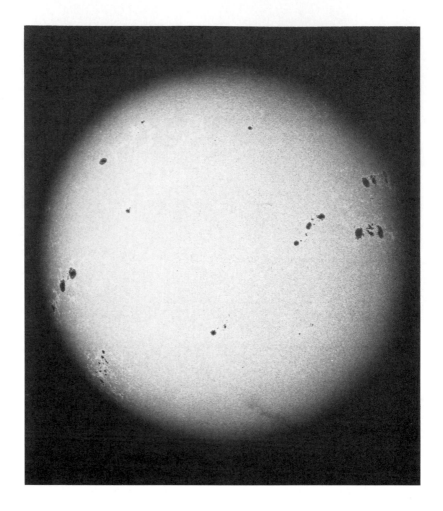

Figure 1-8 Our Star
*The sun is by far the most massive object in the solar system.
Over 99.8 percent of all the matter in the solar system is located in
the sun. Less than 0.2 percent was left over for everything else. An
exceptionally large number of sunspots are seen in this particular
photograph. (Hale Observatories.)*

and meteoroids — could resist these winds and remain in orbit about the sun.

Although the planets would continue to evolve over the next few billion years, the creation of the solar system was finished. Aside from processes such as the cratering of the inner planets, no truly major changes would occur after the T Tauri stage. The T Tauri wind terminated the process of planet creation.

Following the cessation of the T Tauri wind, most of the matter remaining in the solar system was contained in the sun. This is how we find it today. Over 99.8 percent of the mass of the solar system is located in the sun. That leaves less than 0.9 percent for all the planets combined. Comets, asteroids, satellites, and meteoroids together make up less than 0.001 percent of the solar system's mass.

A space traveler whose galactic wanderings brought him or her toward the solar neighborhood would notice only the sun. At first glance, only a dim, dwarf star would be seen. But with careful scrutiny at a close range of less than a light year, Jupiter would be visible. Then Saturn. Only with great difficulty, or at very close range, would the traveler become aware of any of the other planets. Quite literally, the planets are microscopic impurities in the vast cosmic vacuum that surrounds our star.

2

Sun-Scorched
Mercury

Mercury is the closest planet to the sun. The average distance to the sun from Mercury is only 58 million kilometers (36 million miles). For comparison, Earth's orbit is nearly three times as large. This means that the sun, which appears to have an angular diameter of ½° when seen from Earth, has a diameter of nearly 1½° when viewed from Mercury. Because of this threefold increase in apparent diameter, the sun's surface area appears nine times as large. As seen from Mercury, the sun covers an area in the sky nine times as large as it does from Earth. Therefore, compared with Earth, nine times as much sunlight beats down on Mercury's cratered surface. But unlike our planet, Mercury has no protective atmosphere to filter the searing solar rays. Under the blistering noontime sun, any lead and tin would be sweated out of the rocks and collect in glistening puddles or flow across the planet in thin silvery streams. It is one of the hottest places in the solar system, a silent, airless inferno, completely dominated by the huge sun suspended against a jet-black sky.

Mercury's proximity to the sun means that it is extremely difficult to get a clear view of this innermost planet. Indeed, Mercury is always so close to the sun that many professional astronomers have never glimpsed the tiny planet. Although Mercury is often very bright (occasionally it becomes as bright as Sirius, the brightest-appearing star in the sky), its brilliance is lost in the blues and pinks of the dawn or dusk.

As shown in Figure 2-1, Mercury is never more than 28° away from the sun. This means that—at best—Mercury rises in the east 1½ hours before the sun or sets in the west 1½ hours after the sun. The planet is therefore always buried in the glow of twilight.

As also indicated in Figure 2-1, Mercury's orbit is not circular. Although Mercury's orbit looks roughly circular at first glance, it is actually the most highly elliptical orbit of any planet in the solar system, except for Pluto. At perihelion, the distance to the sun is only 46 million kilometers (28½ million miles). But at aphelion, the distance to the sun is nearly 70 million kilometers (nearly 43⅔ million miles). This is a substantial distance variation as Mercury goes around its orbit once every 88 days.

Mercury is the second smallest planet in the solar system. With a diameter of only 4,880 kilometers (3,030 miles), Mercury is

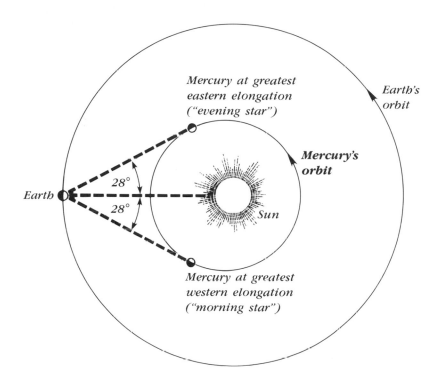

Mercury at greatest
eastern elongation
("evening star")

Earth's
orbit

Mercury's
orbit

Earth

28°

28°

Sun

Mercury at greatest
western elongation
("morning star")

Figure 2-1   The Orbit of Mercury

At an average distance of only 58 million kilometers (36 million miles) from the sun, Mercury takes a mere 88 days to go around its orbit. As viewed from Earth, Mercury can be seen only near times of greatest eastern or western elongation. At greatest western elongation (when the planet is farthest west of the sun in the sky), Mercury rises about 1½ hours before sunrise. At greatest eastern elongation (when the planet is farthest east of the sun in the sky), Mercury sets about 1½ hours after sunset.

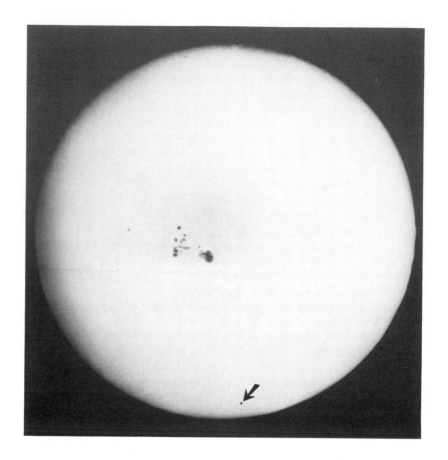

Figure 2-2   A Transit of Mercury
*On rare occasions (roughly a dozen times during a century),
Mercury happens to pass directly between the Earth and the sun.
On such occasions, the tiny planet is seen silhouetted against the
sun's surface. This particular transit occurred on November 14,
1907. Because of the substantial tilt of Mercury's orbit (Mercury's
orbit is inclined by 7° to Earth's orbit), the planet usually passes
somewhat north or south of the sun, and no transit is seen. (Yerkes
Observatory.)*

not much larger than our moon. Indeed, two satellites of Jupiter, one of Saturn, and one of Neptune are all slightly bigger than Mercury.

Mercury's small size and perpetual proximity to the sun make it a very frustrating planet to study from Earth. This sad state of affairs is blatantly obvious in Figure 2-3. These two views are among the finest photographs of Mercury ever taken by earth-based astronomers. Only a few very hazy markings can be distinguished.

Astronomers of the nineteenth century sought to determine Mercury's period of rotation by carefully watching the hazy markings on the planet's surface. In 1889, after seven years of observations, Giovanni Schiaparelli (the same fellow who made the first extensive maps of Mars) announced that "Mercury performs only one rotation during its revolution round the sun." In other words, just as our moon always keeps the same side facing Earth, Mercury allegedly keeps the same side facing the sun. This would mean that one side of Mercury experiences perpetual blistering day while the other side of the planet is forever immersed in frigid night.

Schiaparelli was wrong. But Mercury is so difficult to observe that no one suspected an error until the mid-1960s. Schiaparelli's synchronous rotation of Mercury was taken as gospel; myths and fantasy abounded in the popular and scientific literature.

In 1962, a team of astronomers from the University of Michigan carefully observed Mercury with a radio telescope. By measuring the amount of radio wavelength radiation from the planet, they hoped to determine the surface temperature. Much to their surprise, however, the astronomers detected radio emissions from Mercury's dark, nighttime side. Of course, they expected lots of radio waves from the hot, sunlit portions of the planet. But if the dark side never faced the sun, the temperature of the rocks should be very near absolute zero and no radio waves should be given off.

One obvious explanation of these radio observations is that Mercury does not rotate synchronously, and thus one side of the planet is not subjected to perpetual night. But Schiaparelli's ideas were so ingrained that alternative explanations were sought. Astronomers began talking about an atmosphere around Mercury that could be responsible for transporting heat from the daytime hemisphere to the nighttime hemisphere. It was all a complete fiction.

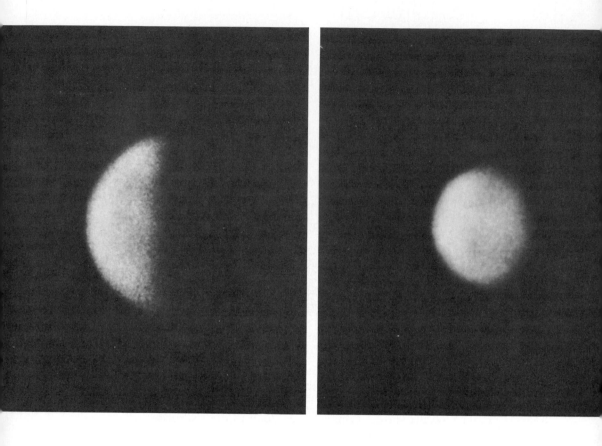

Figure 2-3   Two Views of Mercury
*These two photographs taken in the late 1960s represent the best
views of Mercury that can be obtained from Earth. Only a few
faint dusky markings can be seen. Because of the faintness of these
features, it was very difficult to determine Mercury's period of
rotation. Astronomers now know that a "Mercurian day" (58.65
earth-days) is exactly equal to two-thirds of a "Mercurian year"
(87.97 earth-days). (New Mexico State University Observatory.)*

In 1965, Rolf Dyce and Gordon Pettingill used the giant Arecibo radio telescope in Puerto Rico to bounce radar waves off Mercury. From analyzing the radar echo, they concluded that Mercury does not keep the same side facing the sun. The radar observations clearly indicated that Mercury rotates with a period of 59 days.

Following the announcement of this important discovery, the brilliant Italian astronomer Guiseppe Colombo noticed that Mercury's period of rotation (59 days) appears to be two-thirds of Mercury's period of revolution around the sun (88 days). In other words, instead of Schiaparelli's 1-to-1 relationship between rotation and revolution (in which both the "day" and "year" on Mercury would be equal to 88 earth-days), there appears to be a 2-to-3 relationship. Accordingly, Colombo postulated that a Mercurian day is *exactly* two-thirds of a Mercurian year. This prophetic insight was confirmed ten years later with the flight of Mariner 10. Mercury rotates three times about its axis during every two trips around the sun. A remarkable resonance at the core of the solar system!

Because of Mercury's proximity to the sun, detailed knowledge of the planet could come only from interplanetary spacecrafts. This would be the legacy of Mariner 10. During a few short days in March of 1974, more would be learned about Mercury than the sum of all knowledge collected over the preceding course of human history.

Shortly after midnight on November 4, 1973, a huge Atlas/Centaur lifted laboriously from Pad 36B at Cape Canaveral. A perfect launch that sent Mariner 10 silently gliding to Venus. After skimming past Venus on the morning of February 5, 1974, the spacecraft continued to coast toward its ultimate rendezvous.

The cameras were turned on while the spacecraft was still more than 5 million kilometers from Mercury. At that enormous distance, the view was not much better than in the fuzzy photographs from earth-based telescopes. But over the next few days, as Mariner 10 glided to within 756 kilometers (470 miles) of the airless planet on March 29, 1974, resolution increased to the point that features as small as 150 meters (500 feet) across could be seen. Breathtaking

Figure 2-4 The Flight of Mariner 10
Mariner 10 was launched from Earth on November 4, 1973. As the spacecraft sped past Venus on February 5, 1974, Venus's gravity catapulted Mariner 10 on toward Mercury. The first encounter with Mercury occurred on March 29, 1974. From there, Mariner 10 went into an orbit about the sun that returns the spacecraft to Mercury every two Mercurian years (that is, every 176 days). (NASA.)

vistas unfolded, and unimagined landscapes were revealed as one sharp, crisp picture after the next began to come in. Mercury, at first glance, looks exactly like our moon!

Although picture-taking went on for ten days, the best photographs were taken just before and just after closest approach. A mosaic of the "incoming view" is shown in Figure 2-6. The "outgoing view" is seen in Figure 2-7.

In between the incoming and outgoing views of the planet, there was an interruption of data transmission as the spacecraft passed through Mercury's shadow. As Mariner 10 skimmed over the night side of the planet, instruments onboard the spacecraft measured temperatures as low as −173°C (−280°F). In contrast, noontime temperatures on the planet soar to 427°C (800°F).

In addition to measuring surface temperatures, instruments on Mariner 10 discovered that Mercury has a magnetic field. Although weak by earth standards (Earth's magnetic field is roughly a hundred times as strong as Mercury's), Mercury's magnetic field is much more intense than the fields found around either Venus or Mars.

The existence of a magnetic field around Mercury means that the planet must have an iron core. Only an iron core could support a permanent, planetwide magnetic field. From the overall density of the planet, scientists calculate that Mercury's iron core is 3,600 kilometers (2,200 miles) in diameter. Mercury's huge iron core is the same size as our moon! Since the entire planet is only 4,880 kilometers (3,030 miles) in diameter, the iron core is surrounded by a mantle of rock only 640 kilometers (400 miles) thick.

The existence of a large iron core provided one of the first indications that Mercury, in spite of its lunarlike appearance, is very different from our moon. The moon has neither a magnetic field nor an iron core. Additional differences between the moon and Mercury came to light as scientists scrutinized the Mariner 10 photographs.

Both Mercury and our moon are covered with craters. A typical, heavily cratered region near Mercury's south pole is shown in Figure 2-8. But even in these heavily cratered regions, there are substantial intercrater plains. Relatively smooth areas are usually found separating the craters on Mercury.

Figure 2-5 Mercury's Lunarlike Landscape
*Like our moon, Mercury is covered with craters. This view of the
planet's northern limb was taken from a distance of 77,800
kilometers (48,300 miles). A cliff called Victoria Scarp is seen very
near the limb. The distance across the bottom of the picture is
nearly 600 kilometers (almost 400 miles). (NASA; JPL.)*

This is very unlike our moon. On the moon, there are virtually no intercrater plains. The densely cratered lunar regions (called the "lunar highlands") consist of overlapping craters with no room for any flat areas in between.

It seems very likely that the difference in surface gravity between the moon and Mercury is responsible for the differences in the appearance of the cratered regions. The moon and Mercury are virtually the same size. But, because of its huge iron core, Mercury is more massive than the moon. Scientists usually express the masses of planets or satellites in terms of Earth's mass (rather than in millions or trillions of tons or pounds). Thus, Mercury's mass is equal to $^1/_{18}$ of Earth's mass, while our moon's mass measures only $^1/_{81}$ of Earth's mass. Since Mercury is 4½ times as massive as the moon, things weigh more on Mercury. A 100-pound child from Earth would weigh only 16 pounds on the moon but would tip the scales at 38 pounds on Mercury. Scientists therefore say that the surface gravity on the moon is 0.16, while the surface gravity on Mercury is 0.38. (The Earth's surface gravity is taken to be 1.00.)

The craters on Mercury and the moon were all formed by the impact of meteoroids, asteroids, and other planetesimals in the distant past. During a violent collision with one of these interplanetary rocks, large quantities of material (called "ejecta") were blasted out of the target area and scattered in all directions. But Mercury's surface gravity is 2⅓ times as strong as the moon's. Because of this stronger gravity, ejecta from a crater-forming impact on Mercury did not travel as far as ejecta from a comparable impact on the moon. In fact, ejecta from an impact on Mercury cover an area only one-sixth as large as the area covered on the moon. Consequently, secondary impact craters on Mercury are closely clustered around the primary impact crater, but on the moon they are spread over an area six times as large. This means that the underlying precrater plains on Mercury were not so easily obliterated. Indeed, the ancient Mercurian plains are still visible between the craters.

While carefully examining the Mariner 10 photographs, geologists noticed another significant difference between Mercury and the moon. All around the planet there are shallow scalloped cliffs. These cliffs, technically called "lobate scarps," are typically 1 or 2 kilometers high and several hundred kilometers long.

Figure 2-6 The "Incoming Hemisphere"
This mosaic was constructed from individual photographs taken just before closest approach. The landscape is dominated by numerous craters. Basins and plains are also easily seen. Just after taking these photographs, Mariner 10 coasted into Mercury's shadow and passed within 756 kilometers (470 miles) of the planet's surface. (NASA; JPL.)

Figure 2-7 The "Outgoing Hemisphere"
This mosaic was constructed from individual photographs taken about two hours after the closest approach, as the spacecraft sped away from the planet. The landscape is dominated by larger plains and fewer craters than were seen in the "incoming" view. A huge basin, the Caloris Basin, is located near the middle of the disk, half emersed in the night shadows. (NASA; JPL.)

Figure 2-8   The Densely Cratered South Polar Region
This high resolution view shows a heavily cratered region near Mercury's south pole. Features as small as 2 kilometers (1¼ miles) across can be seen. The south pole is just beyond the bottom of the picture. The dark-rimmed, bright-rayed crater near the top of the picture is 67 kilometers (42 miles) in diameter. This mosaic was taken during the spacecraft's second flyby of the planet on September 21, 1974. (NASA; JPL.)

Scarps are difficult to notice at first glance. A nearly edge-on view of a scarp is seen at the limb of the planet in Figure 2-5. It is believed that these scarps are gigantic wrinkles that formed in Mercury's crust as the planet's huge iron core cooled and contracted. No comparable features are seen on the moon.

While it is sometimes difficult for the untrained eye to see scarps, there is no mistaking the Caloris Basin. The Caloris Basin is the largest single feature viewed from Mariner 10. As shown in Figure 2-9, the basin measures 1,300 kilometers (800 miles) in diameter and is ringed with mountains that rise 2 kilometers (6,500 feet) above the surrounding plains. Unfortunately, half of this fascinating feature was hidden in the night shadows during the spacecraft's flyby.

The Caloris Basin derives its name from the fact that, every two Mercurian years, it is at the subsolar point when the planet is at perihelion. In other words, every 176 days, when Mercury is nearest the sun, it is high noon at the Caloris Basin. During every other orbit of the sun, the Caloris (from the Latin word meaning "hot") Basin is the hottest place on the planet.

The Caloris Basin is obviously a vast impact feature. Near the end of the cratering epoch some 3 to 4 billion years ago, a large asteroid — perhaps the largest ever to strike Mercury — crashed into the planet. Unlike earlier impacts that merely peppered the Mercurian surface with pockmarks, this violent collision broke through to the molten interior. Large quantities of lava welled up and filled in the enormous crater. Waves on the sea of molten rock were preserved for eternity as the lava cooled and solidified.

It seems that the planet-shaking impact that formed the Caloris Basin had a significant effect on the terrain at other locations around Mercury. Antipodal to the Caloris Basin (that is, exactly on the opposite side of the planet from the Caloris Basin), there is an extensive region of peculiar rippled terrain. This terrain, shown in Figure 2-10, consists of thousands of closely spaced, lumpy-looking hills with heights ranging from roughly ¼ to 2 kilometers. It is reasonable to suppose that powerful seismic waves from the Caloris impact were focused as they passed through the planet. When the focused waves arrived at the other side of the planet, the ground vibrated and shook with such violence that thousands of mile-high

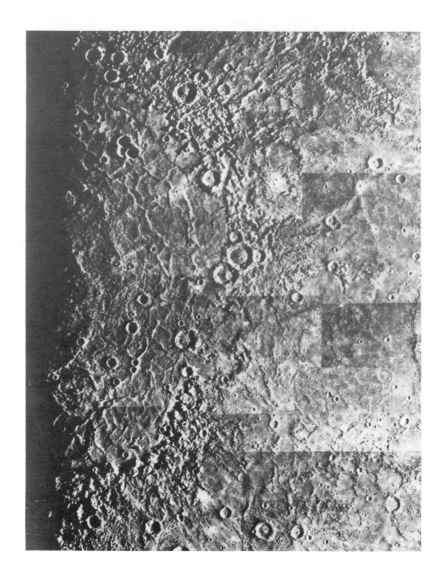

Figure 2-9   The Caloris Basin
This vast basin, half hidden by the planet's night side, was the largest single feature viewed from Mariner 10. The outer edge of the basin is outlined by a ring of mountains 1,300 kilometers (800 miles) in diameter. These mountains rise to elevations of 2 kilometers (6,500 feet) above the wrinkled basin floor. (NASA; JPL.)

mountains were pushed up in only a few seconds. This must have been the single most cataclysmic event in the planet's entire history.

On March 30, 1974, with the planet receding rapidly in the distance, the cameras onboard Mariner 10 were shut off. Only 24 hours earlier, human eyes had seen hundreds upon hundreds of spectacular photographs of an unimagined, desolate landscape never before viewed by any creature in the universe. And then, during the early days of April, all the craters, the plains, the hills, and the basins merged and faded into indistinct fuzzy features that are seen from afar. But the mission was not over.

Four years earlier, while the Mariner 10 mission was still being planned, Giuseppe Colombo took great interest in the orbit of the spacecraft around the sun after the Mercury flyby. He found that Mariner 10 would ultimately go into a highly elliptical orbit, revolving about the sun once every 176 days. But that is exactly two Mercurian years! Mariner 10 would come back to Mercury every 176 days. There would be a second chance. And a third.

Mariner 10 flew past Mercury for the second time on September 21, 1974. Almost 2,000 more photographs were obtained, including those shown in Figure 2-8. And during the afternoon of March 16, 1975, Mariner 10 once again skimmed over the planet's surface—this time at the hair-raising distance of only 300 kilometers—again returning numerous photographs. But no new features were seen.

Mariner 10 comes back to Mercury every two Mercurian years. But recall that two Mercurian years is exactly equal to three Mercurian days. Thus, each time Mariner 10 comes back to Mercury, the planet has rotated exactly three times. This means that the same craters and plains are facing the sun during each flyby. The view at each flyby is always essentially the same.

Mariner 10 has seen half the planet. After the third flyby, insufficient fuel remained to keep the spacecraft from tumbling aimlessly. But Mariner 10 still returns to Mercury every 176 days. Every two Mercurian years, the same craters, plains, and basins are exposed to blind mechanical eyes as the spacecraft coasts helplessly along its perpetual orbit.

Mariner 10 has photographed half the planet. It cost the American people $98 million to view half a world never before

Figure 2-10   Peculiar Terrain

*Antipodal to the Caloris Basin (that is, on the opposite side of the
planet from the Caloris Basin), there is an unusual rippled or
wrinkled terrain. This peculiar terrain covers 500,000 square
kilometers (193,000 square miles), an area nearly twice the size of
Colorado. Powerful seismic waves from the Caloris impact may
have caused the corrugated appearance of this part of the planet.
(NASA; JPL.)*

seen by human eyes. This total cost is the same as 50 cents per person—not enough to buy a single gallon of gasoline! Yet no return missions to Mercury have been funded. Unless the American people become willing to accept a financial burden equal to a one-time purchase of one gallon of gasoline or one pack of cigarettes, the other half of the planet shall remain shrouded in darkness and mystery.

3
Cloud-Covered Venus

Venus is often one of the brightest objects in the sky. At greatest brilliancy, Venus is ten times as luminous as Sirius, the brightest-appearing star. Only the sun and the moon manage to outshine this planet.

Venus is bright because the entire planet is surrounded by shimmering yellowish-white clouds and also because it is near the sun. Venus is the second planet from the sun. At a distance of only 108 million kilometers (67 million miles) from the sun, Venus is bathed in twice as much sunlight as Earth. Venus's thick cloud cover easily reflects the majority of all this sunlight, thereby making the planet one of the most dazzling objects in the heavens.

Because Venus is closer to the sun than Earth is, Venus can be seen only in the west after sunset (as the so-called evening star) or in the east before sunrise (as the so-called morning star). This situation is similar to the case of Mercury. But since Venus's orbit is twice as big as Mercury's, Venus can appear at conveniently large distances from the sun. As shown in Figure 3-1, Venus can be seen at distances of up to 47° from the sun in the sky. This means that Venus can rise as early as three hours before the sun or set as late as three hours after the sun. At these favorable times, earth-based observers have no difficulty viewing the planet. While many professional astronomers have never glimpsed Mercury, most people around the world have seen Venus — if only to wonder what that dazzling object could be.

Venus once played a crucial role in the history of astronomy. In the early 1600s, Galileo Galilei heard rumors of an extraordinary device that consisted of lenses mounted at the ends of a hollow tube. When one looked through the contraption, distant objects appeared closer and larger than usual. Galileo promptly set about the business of building one.

Galileo did not invent the telescope. He was simply the first person to point it toward the sky. His observations and discoveries resulted in the first new, fresh revelations about the universe in over a thousand years. He found craters on the moon, sunspots on the sun, and four moons orbiting Jupiter. He also discovered that Venus, like our moon, goes through phases.

While observing Venus over many months, Galileo noticed that the phase of Venus is correlated with the apparent size of the

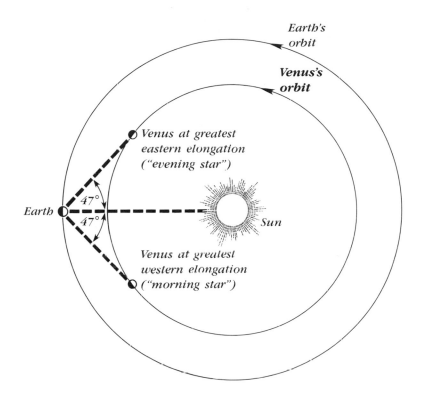

Figure 3-1 The Orbit of Venus
At an average distance of 108 million kilometers (67 million miles) from the sun, Venus takes 225 days to go around its orbit. Earth-based observers have no trouble seeing the planet near greatest eastern or greatest western elongation. For weeks around the time of greatest western elongation, Venus completely dominates the predawn sky. For weeks around the time of greatest eastern elongation, Venus easily outshines all the stars in the western sky after sunset.

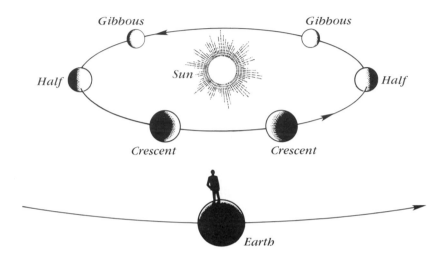

Figure 3-2   The Phases of Venus
Earth-based observers often see substantial portions of Venus's dark, nighttime side. When the planet is near Earth, Venus appears large and exhibits a crescent phase. When the planet is far form Earth, Venus appears small and exhibits a gibbous phase.

planet as viewed through his telescope. When Venus exhibits a crescent phase, the planet appears very large. But months later, when Venus exhibits a gibbous phase, the planet's disk is quite small. As shown in Figure 3-2, this relationship between the size and phase of Venus clearly demonstrates that the planet goes around the sun. For the first time in history, humanity had observational evidence that the heretical ideas of Copernicus and Kepler might be right after all. The sun and not the earth is indeed at the center of the universe. For this and similar blasphemies, Galileo was tried, convicted, and sentenced to perpetual house arrest by the Inquisition.

During the centuries that followed Galileo, it became apparent that Venus and Earth had several striking similarities. Both planets have nearly the same size, the same mass, the same overall density, and the same surface gravity. They both even have atmospheres. The only obvious difference is that Earth has a moon and Venus does not. In addition, Venus is perpetually and completely cloud-covered, while Earth is only partly cloud-covered. Thus, Venus became known as Earth's twin sister; some people even speculated about the possibility of abundant life beneath the Venusian clouds. Nothing could be farther from the truth. The surface of Venus is a blistering inferno, hotter and more vicious than the sun-scorched craters of Mercury.

Until recently, little was known about Mercury because the tiny planet is so close to the sun and therefore difficult to observe from Earth. While Venus is farther from the sun and thus easily observed from Earth, Venus also remained enshrouded in mystery. Earth-based observers can only see the shimmering cloud-tops.

The first attempt to penetrate the thick Venusian cloud cover came in the early 1960s. Scientists in the United States and the Soviet Union used their satellite-tracking antennas to beam radar waves at Venus. Radar waves easily penetrated the clouds and were reflected off the planet's surface. Analysis of the radar echo revealed that Venus rotates very slowly, once every 243 days, retrograde. In other words, as seen from Venus, the sun rises in the west and sets in the east.

This retrograde (or backward) period of Venus is measured with respect to the *stars*. Every 243 days, a person on Venus would see the same configuration of stars in the sky. But during that time,

Figure 3-3   Venus Seen from Earth
*Venus's surface is perpetually hidden from view by a thick,
yellowish-white cloud cover. By bouncing radar waves off the
planet's surface, scientists have discovered that Venus rotates
backward. As seen from the Venusian surface, the sun would appear
to rise in the west and set in the east. (Hale Observatories.)*

Venus has moved a substantial distance along its orbit. Although 243 days must elapse before the same stars are seen in the same positions again, a person on Venus would have witnessed a couple of sunrises and sunsets. A "Venusian day" (that is, the period of rotation measured with respect to the *sun*) lasts for only 116.8 earth-days. Thus, as viewed from the Venusian surface, 116.8 days elapse between one high noon and the next.

Every 584 days, Venus passes between the Earth and the sun. Actually, it is very rare that Venus is seen silhouetted against the solar surface (a so-called transit) because Venus's orbit is tilted slightly with respect to Earth's orbit. The last transit of Venus occurred in 1882 and the next is scheduled for June 8, 2004 — none at all during the twentieth century. But nevertheless, as viewed from Earth, Venus passes slightly north or south of the sun every 584 days. Each of these passes, when the night side of Venus is exposed to earth-based observers, is called an "inferior conjunction."

It is striking to notice that 584 days is exactly equal to 5 times 116.8 days. In other words, the time between successive inferior conjunctions is exactly equal to five Venusian days. Since exactly five Venusian days elapse between inferior conjunctions, the same night side of Venus is always exposed to Earth. Surprisingly, therefore, Venus's backward rotation is locked into Earth's revolution about the sun. If we could see through Venus's clouds, we would view exactly the same nighttime landscape every time Venus passed between us and the sun. Another extraordinary resonance in the solar system!

This resonance, in which Venus's rotation is tied to Earth's revolution, has been demonstrated from radar observations. While we cannot see through the cloud cover, radar waves easily penetrate the thick atmosphere and are reflected from the planet's surface. From analyzing the radar echos received back at Earth, Richard M. Goldstein has been able to deduce the existence of several large craters on the Venusian surface. At each and every inferior conjunction, when Venus is nearest Earth, these same craters are always exposed to our radar beams.

Because of an eternal cloud cover, detailed information about Venus had to come from interplanetary space flights. For this reason, Venus has been one of the prime targets of both the Soviet and

American space programs. Indeed, just about as many spacecrafts have been sent to Venus as have been launched toward Mars.

The Americans and the Soviets adopted very different approaches to the problem of probing Venus. The American tactic was to fly spacecrafts past the planet. During the encounter, instruments would remotely measure a variety of atmospheric conditions. The Soviet plan was to plunge directly into the Venusian atmosphere and measure conditions as the vehicle descended to the surface. The Americans used delicate, compact, lightweight spacecrafts that are still in orbit about the sun. The Russians ended up sending monstrous contraptions covered with cables and shock absorbers and steel plating and parachutes.

Mariner 2 was the first spacecraft to undertake a successful mission to Venus (two earlier attempts, Venera 1 and Mariner 1, malfunctioned). Indeed, the flight of Mariner 2 in 1962 was the first successful flyby of any planet—a sorely needed morale boost for the fledgling American space program. Temperatures of 400°C (750°F) were found to exist in the Venusian atmosphere.

These high temperatures confirmed earlier suspicions that Venus is indeed a very hot place. The primary mechanism that keeps Venus so warm is called the "greenhouse effect." To understand this effect, imagine an automobile parked by the curb on a sunny day. Further imagine that all the windows are rolled up and the car simply sits in the sunshine for a few hours. It is a common experience that the temperature inside the car can get much higher than the outside air temperature. The reason is that visible sunlight easily enters the car through the glass windows. Some of the sunlight is reflected back out, but the rest is absorbed inside the car by objects (the seats, for example) that begin to heat up. These warm objects reradiate the energy they absorbed, but at much longer infrared wavelengths. Infrared light cannot pass through glass, and thus the energy remains trapped inside the car. And the car's interior gets hot. Greenhouses stay warm in this fashion.

In exactly the same way, sunlight is absorbed and trapped by the Venusian clouds. This is why the atmosphere of Venus is so warm. But no one imagined that it could be hotter than Merucry.

Much improved data about the Venusian atmosphere came from the flights of Mariner 5, Venera 4, Venera 5, and Venera 6 in the

Figure 3-4 Venus Seen from Space
*This spactacular close-up view of Venus was photographed by
Mariner 10 on February 6, 1974, at a distance of 720,000
kilometers (450,000 miles). The top of the cloud cover is at an
altitude of 65 kilometers (40 miles) above the planet's surface.
(NASA; JPL.)*

late 1960s. By this time, it was clear that the high temperatures on the planet are accompanied by an enormous, crushing atmospheric pressure. And while the Americans concentrated on manned lunar exploration, the Soviets built heavier and sturdier spacecrafts to withstand these harsh conditions on Venus. Their efforts culminated with the successful descents of Venera 7 and Venera 8 in the early 1970s. Indeed, Venera 8 discovered the *bottom* of the Venusian cloud cover at an altitude of 35 kilometers (22 miles) above the planet's surface. The cloud cover does not extend down to the ground. Several of these later Venera spacecrafts apparently succeeded in transmitting data from the planet's surface.

By piecing together all the data from the Mariner and Venera flights, scientists have been able to assemble a fairly complete picture of Venus's atmosphere. The temperature in the Venusian atmosphere is shown in Figure 3-5. In the cloud-tops at an elevation of about 65 kilometers (40 miles) above the planet's surface, the temperature is a cool −50°C (−60°F). Descending into the clouds, we find that the temperature rises dramatically. In fact, at the surface, the temperature is a blistering 475°C (890°F). For comparison, recall that typical noontime temperatures on Mercury measure only 430°C (800°F). But unlike the case of airless Mercury, the night side of Venus never manages to cool off. The dense atmosphere keeps heat from radiating into space during the long Venusian night. It is so incredibly hot on Venus that during the night on the dark side of the planet, rocks glow with a dull red light, like the hot coils of an electric stove.

In addition to high temperatures, the atmospheric pressure on Venus is also enormous; it is shown in Figure 3-6. Scientists often prefer to measure pressure in "atmospheres," where "1 atmosphere" is equal to the average air pressure here on Earth at sea level (nearly 15 pounds per square inch). As shown in Figure 3-6, a pressure of 1 atmosphere on Venus is encountered at an elevation of 50 kilometers above the planet's surface. Down on the surface, the pressure has risen to a crushing 90 atmospheres (⅔ ton per square inch). For comparison, you would have to descend to a depth of 3,000 feet into the ocean to encounter similar pressures here on Earth.

Finally, Mariner and Venera flights have revealed that Venus's atmosphere is composed almost entirely of carbon dioxide. Carbon dioxide accounts for 97 percent of the Venusian atmosphere. Of the remainder, 2 percent is nitrogen, which leaves only 1 percent for all other gases combined. For comparison, Earth's atmosphere consists of nearly 80 percent nitrogen, 20 percent oxygen, and less than one-tenth percent carbon dioxide.

Some of the best and most complete information about Venus came from Soviet and American flights during the mid-1970s. First of all, Mariner 10 flew past Venus on February 5, 1974, while on the way to Mercury. During the encounter with Venus, numerous photographs were taken that reveal a very turbulent atmosphere. A planetwide view is shown in Figure 3-4, while Figure 3-7 gives a close-up of the cloud structure.

The Mariner 10 photographs revealed a phenomenon that had been suspected from earth-based observations. As shown in Figure 3-8, the atmosphere rotates about the planet very rapidly. From watching features in the clouds, we realize that it takes only four days for the atmosphere to go completely around the planet. Of course, the planet itself rotates much more slowly—once every 243 days. This means that the wind velocity must be huge, typically 100 meters per second (225 miles per hour). The rotation of both the atmosphere and the planet are retrograde.

Since Venus's atmosphere is so thick and moves so rapidly across the planet, many scientists once believed that photography from the planet's surface would be a waste of time and effort. Perhaps the thick atmosphere keeps out the sunlight so that the surface is in perpetual darkness. After all, it is essentially pitch black at 3,000 feet below the ocean's surface here on Earth. And even if some sunlight leaked through, you would only see a sandblasted, smooth surface because of the excessive wind speeds. Or at least this is what quite a few people thought.

In October 1975, the Soviet Union scored an impressive double success with the soft landings of Venera 9 and Venera 10. Both spacecrafts survived on the surface long enough to send back panoramas of the landing sites. To everyone's surprise, angular rocks could be seen lying all over the place. As shown in Figure 3-9, the Venera 9 site appears to be somewhat more boulder-strewn than the Venera 10 site. In addition to measuring surface temperatures and

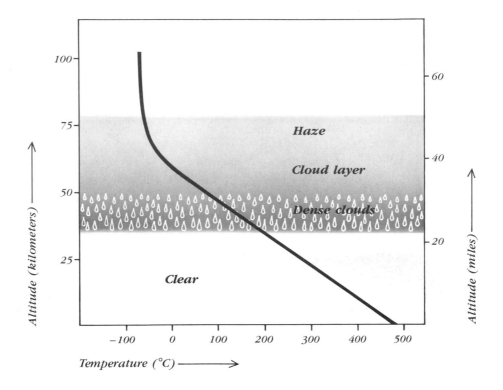

Figure 3-5   Temperature in the Venusian Atmosphere
*Combined data from Mariner and Venera missions give this
temperature profile of Venus's atmosphere. Temperature at the
cloud-tops is about −50°C (−60°F). Below the cloud-tops, the
temperature rises dramatically with decreasing altitude. Venus's
surface temperature is a blistering 475°C (890°F).*

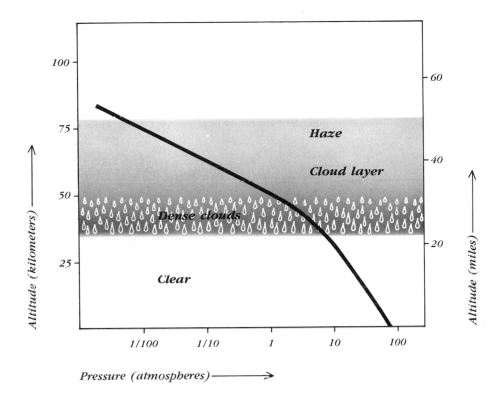

Figure 3-6 Pressure in the Venusian Atmosphere
Combined data from Mariner and Venera missions give this pressure profile of Venus's atmosphere. (Pressure is measured in atmospheres. One atmosphere equals the air pressure on Earth at sea level.) The atmospheric pressure at ground level on Venus is a crushing 90 atmospheres (1,300 pounds per square inch).

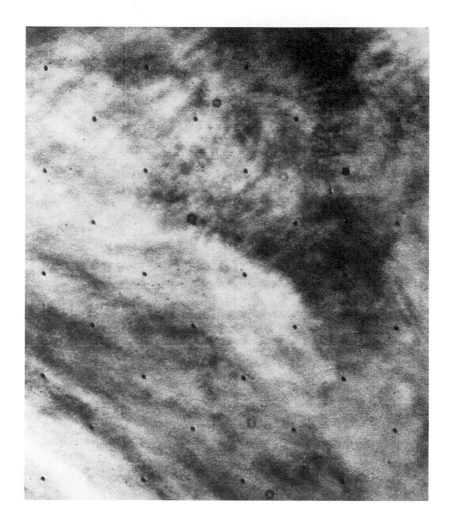

Figure 3-7   A Close-up of the Venusian Clouds
*On Febuary 5, 1974, Mariner 10 glided past Venus at an altitude of
only 5,800 kilometers (3,600 miles) above the cloud-tops. This
photograph was taken shortly after closest approach. Details as
small as 7 kilometers (4 miles) in size can be seen. (NASA; JPL.)*

pressures, the Soviet spacecrafts found a surprisingly low wind speed—only 3½ meters per second (8 miles per hour). This low wind velocity at the Venusian surface means that there must be enormous shear in the clouds where the transition from high speed to low speed occurs.

At first glance, an alien visitor to our solar system might think that Venus and Earth could have a lot in common. They have nearly identical sizes, masses, densities, and so forth. And they have orbits alongside each other in the same part of the solar system. Thus, they must have formed from essentially the same primordial material. But close observation soon reveals that the two planets are incredibly different. What could have happened that made the planets so dissimilar?

It is perhaps surprising to realize that there are enormous amounts of carbon dioxide here on Earth. But this terrestrial carbon dioxide is chemically bound in rocks as calcium carbonate. Certain rocks can react with carbon dioxide in the air to form calcium carbonate much as oxygen in the air reacts with iron to form rust. More importantly, marine animals have been removing carbon dioxide from seawater (and thus ultimately from the atmosphere) for hundreds of millions of years. The resulting seashells that contain the carbon dioxide have been transformed into limestone and marble over the ages. And, of course, plants have been removing carbon dioxide from the air by photosynthesis for billions of years, thereby enriching the atmosphere with oxygen. The carbon from this carbon dioxide is contained in deposits of coal and oil.

Carbon dioxide is an excellent gas to support a runaway greenhouse effect. As on Venus, there is no stopping the effect once it gets started. On Venus, it was never cool enough for chemical or biological reactions to remove the carbon dioxide from the atmosphere. Earth may have had essentially the same amount of carbon dioxide as Venus. But on Earth—only 41 million kilometers further from the sun than Venus—it was never quite warm enough to start a greenhouse effect, and thus methods for extracting carbon dioxide from the air could get under way. Those few extra millions of kilometers from the sun literally made the difference between life and death.

a

2d   0h

Figure 3-8   The Rotation of Venus's Atmosphere
*These three views were photographed at seven-hour intervals
beginning two days after Mariner 10 flew past Venus. By following
cloud patterns across the planet (see arrow), we see that Venus's
atmosphere rotates around the planet once every four days.
(NASA; JPL.)*

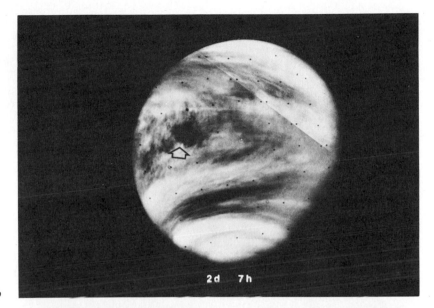

b

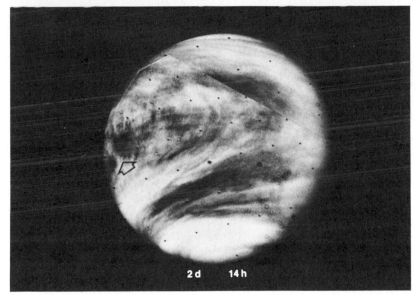

c

BEHEPA-9

BEHEPA-10

Figure 3-9 Venus's Rocky Surface
*In October 1975, two Soviet spacecrafts (Venera 9 and Venera 10)
soft-landed on Venus and sent back these panoramas of the planet's
surface. The view from Venera 9 (upper photograph) showed sharp
rocks measuring 12 to 16 inches across. The Venera 10 site (lower
photograph) was apparently smoother. (National Space Science
Data Center.)*

Five billion years from now, major changes will occur in the structure of the sun.* In response to the depletion of nuclear fuel, the sun will expand enormously. As the sun's surface draws near our planet, the oceans will boil and the rocks will begin to give up their carbon dioxide. Earth will develop an oppressive carbon dioxide atmosphere beneath a perpetual cloud cover. Perhaps, as we observe Venus, we are really seeing the future of our own planet.

---

*For a detailed discussion of stellar evolution, the reader is referred to *Stars and Nebulas* (W. H. Freeman and Company, San Francisco, 1978) by the author of this book.

4

Our Living Earth

Of all the objects that orbit the sun, we are most familiar with our own planet: Earth. We live on its surface. We drink its water. And we breathe its atmosphere. It is obvious, therefore, that more is known about the third planet from the sun than about any other object in the universe.

To alien space travelers, it would be immediately apparent that Earth is an active and dynamic planet. Swirling cloud patterns in Earth's atmosphere would greet the aliens as their spacecraft approached the inner regions of the solar system. From afar, Earth would probably look more interesting than either of its neighbors. After all, Venus is perpetually surrounded by a thick, unbroken cloud cover that never permits a glimpse of its surface. And aside from an occasional planetwide dust storm, Mars's atmosphere hardly ever exhibits any dramatic activty. But as seen from space, Earth continuously displays an intricate ballet of blue-and-white filigree. The delicate shapes and swirls constantly change and shift, come and go, in an unending response to variations in atmospheric pressure, temperature, and moisture content around the planet.

With closer scrutiny and analysis, the alien space travelers might be very puzzled to find that Earth's atmosphere is vastly different from the atmospheres of either of its neighbors. As shown in the table, the atmospheres of both Venus and Mars consist almost entirely of carbon dioxide. But carbon dioxide accounts for only 0.03 percent of Earth's atmosphere. Most of Earth's atmosphere consists of a 4-to-1 mixture of nitrogen and oxygen. And these two gases are found only in very small amounts on Venus and Mars.

How can this be? Why is Earth with its nitrogen-oxygen atmosphere bracketed by two planets each possessing nearly pure carbon dioxide atmospheres? Only after landing on our planet and examining geological records would the alien visitors realize that living organisms and biological processes have been active on Earth for at least the past 3 billion years. The relentless action of biological processes such as photosynthesis is largely responsible for the chemical composition of Earth's atmosphere today. In fact, alien visitors might deliberately single out Earth because of the high percentage of oxygen in its atmosphere. As far as anyone knows, large amounts of oxygen in a planetary atmosphere can arise *only* as a direct result of

Figure 4-1 Our Planet
*Astronauts often report that Earth is the most invitingly beautiful
object they can see in the sky. Our blue-and-white world is the
largest of the inner planets in the solar system. (NASA.)*

The Chemical Composition of Three Planetary Atmospheres

	Venus	Earth	Mars
Nitrogen	2%	78%	3%
Oxygen	Almost zero	21%	Almost zero
Carbon dioxide	97%	Almost zero	95%
Other gases	1%	1%	2%

biological activity. Lots of oxygen surrounding a planet would therefore seem to be a sure sign of life.

If alien visitors decided to try to land their spacecraft on
Earth, they would soon discover a second important way in which
our planet differs radically from either of its neighbors. Earth is wet.
Very wet. Nearly 71 percent of Earth's surface is covered with water.
Indeed, diligent alien visitors might send back thousands of photographs such as Figure 4-3 to their comrades. Such pictures would be
correctly advertised as random close-up views of Earth's surface. By
contrast, Venus and Mars are extremely arid. No place on Earth is as
dry as the surfaces of Venus and Mars. By Venusian or Martian standards, our Sahara desert is a sopping wet swamp.

Although more than two-thirds of our planet is covered by
oceans, Earth is mostly composed of solid rock. Rocks that you find
beneath your feet are typical specimens of Earth's outermost layer, or
crust. All of these rocks fall into one of three categories, depending
on how they were formed.

Igneous rocks are rocks that have cooled from a molten state.
You can see igneous rocks forming as lava runs down the sides of a
volcano. The bedrock under the continents and the oceans is igneous. Roughly two-thirds of all the rocks in Earth's crust are igneous. A
familiar igneous rock called granite is shown in Figure 4-4a.

Sedimentary rocks are rocks that have formed or been depos-
ited by the action of wind, water, or ice. Some sedimentary rocks are
formed when separate rock fragments are cemented together.

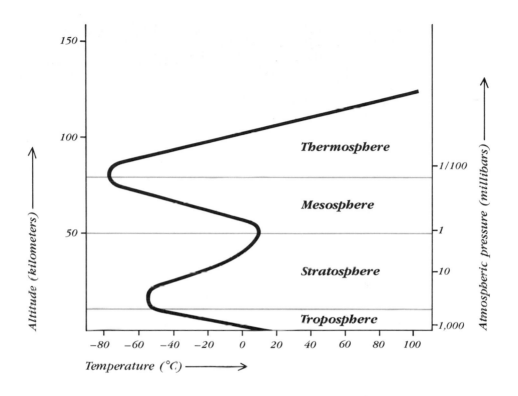

Figure 4-2 The Structure of Earth's Atmosphere
*Atmospheric pressure above Earth's surface decreases steadily with
increasing altitude (see scale on right). In contrast, air temperature
varies in a complicated fashion with altitude. As shown in the
diagram, four distinct regions in Earth's atmosphere are recognized.*

Figure 4-3 A Close-Up View of Earth's Surface
More than two-thirds of Earth's surface is covered with water. By contrast, there is virtually no water on Mercury, Venus, or Mars. (Courtesy of Lynne Ericksson.)

Sandstone, shown in Figure 4-4b, is a good example. Other sedimentary rocks can form as materials once dissolved in water precipitate out at the bottom of a lake or ocean. Limestone and chalk are familiar sedimentary rocks that form by precipitation.

Finally, whenever an igneous or sedimentary rock is subjected to extreme pressures and temperatures beneath Earth's surface, the rock is changed, or metamorphosed. The resulting specimen is called a *metamorphic rock.* For example, when limestone is subjected to heat and pressure, it becomes marble. Marble, shown in Figure 4-4c, is a well-known metamorphic rock.

It is apparent that rocks found in Earth's crust cannot be representative of the planet as a whole. From knowing both the mass and the size of Earth, it is easy to calculate our planet's average density. The average density of Earth is 5½ grams per cubic centimeter. But the average density of a typical crustal rock is only 2½ grams per cubic centimeter. This means that Earth's interior must consist of a very dense substance. Of all the heavy elements, iron is by far the most abundant in the universe. We therefore conclude that Earth, like Mercury and Venus, must have an iron core.

In many respects, Earth's interior is as hidden and remote from us as the most distant galaxies and quasars in the universe. The deepest wells and mine shafts only scratch Earth's surface. Material from the bottom of ocean trenches or lava that gushes up in a volcano is still quite similar to crustal rocks. Earth's interior has never been tapped. The internal structure of our planet can only be deduced or inferred indirectly.

Earthquakes provide the most important clues to the nature of Earth's interior. Occasionally and almost without warning, huge stresses that have built up in our planet's crust are released with a sudden and violent vibratory motion of Earth's surface. As the resulting seismic waves travel throughout the planet, they are bent and altered by the material through which they pass. By studying these earthquake waves at seismographic stations around the world, and by fitting all the data together, we can deduce some important characteristics of Earth's interior.

From analyses of seismic waves generated by earthquakes, it is clear that Earth's interior can be divided into three main regions: the

a

Figure 4-4a   An Igneous Rock
*Igneous rocks are rocks that form from a molten state. Granite,
shown here, is a common coarse-grained igneous rock. (Ward's
Natural Science Establishment, Inc.)*

Figure 4-4b   A Sedimentary Rock
*Sedimentary rocks form either from the cementing of small rock
fragments (such as sandstone, shown here) or from precipitation
from solution in water (such as limestone). (Ward's Natural Science
Establishment, Inc.)*

Figure 4-4c   A Metamorphic Rock
*Metamorphic rocks are rocks that have been changed by high
pressure and temperature beneath Earth's surface. For example,
when limestone is subjected to heat and pressure, it becomes
marble, shown here. (Ward's Natural Science Establishment, Inc.)*

b

c

crust, the mantle, and the core. The *crust,* with which we are all personally familiar, is very thin. It is thinnest under the oceans, where it may extend to a depth of only 5 kilometers. On the continents, the crust is typically 35 kilometers thick. In some places, the crust may extend to depths of 70 kilometers underneath mountain ranges.

Immediately beneath Earth's crust is a thick layer called the *mantle.* The mantle is composed of iron-rich igneous rock and extends down to a depth of 2,900 kilometers (1,800 miles). As shown in Figure 4-5, the mantle is sandwiched between the crust and the *core.* Earth's iron core measures 7,940 kilometers (4,320 miles) across. Although this is more than half of the planet's total diameter, the core constitutes only one-sixth of Earth's total volume.

Detailed analysis of seismic waves that pass directly through the planet reveal that Earth's core actually consists of two parts: a liquid outer layer surrounding a solid center. As diagramed in Figure 4-5, the outer liquid layer has a thickness of 2,220 kilometers (1,380 miles), while the inner solid core has a diameter of 2,510 kilometers (1,560 miles).

This curious situation — a liquid layer sandwiched between a solid inner core and a solid mantle — is directly related to temperatures and pressures deep within the planet. The temperature at Earth's center is believed to be about 4,200°C (7,600°F). Normally this would be more than hot enough to melt almost anything. But the pressures at Earth's center are so incredibly enormous (more than 3½ million atmospheres) that rocks are prevented from melting. In spite of the high temperature, huge pressures at Earth's center ensure that the inner core is solid. Similarly, in the mantle, the pressure is sufficiently high and the temperature sufficiently low to ensure that the rocks do not melt. But in the liquid layer, these conditions are reversed. High pressures are outstripped by higher temperatures in this region and the rock is liquefied.

The liquid portions of Earth's core are believed to be responsible for our planet's magnetic field. Like a giant dynamo, electric currents in the liquid core produce a planetwide magnetic field. This magnetic field extends tens of thousands of kilometers out into space, where it interacts strongly with the solar wind. As mentioned in Chapter 1, the sun constantly sheds high-speed-charged particles

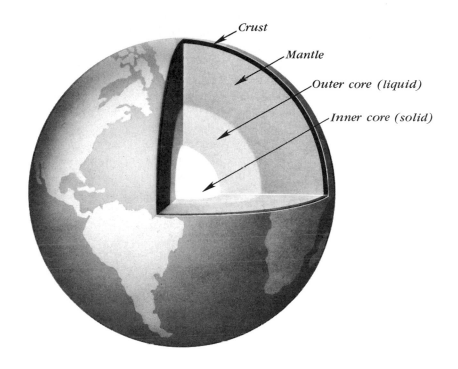

Crust

Mantle

Outer core (liquid)

Inner core (solid)

Figure 4-5   The Structure of Earth's Interior

Earth's interior consists of three basic parts: a thin crust, a mantle, and a core. The iron core is divided into an outer liquid core surrounding an inner solid core. Many important features of Earth's interior are deduced from studying seismic waves caused by earthquakes.

(mostly protons and electrons) in a phenomenon called the *solar wind*. As the supersonic solar wind strikes the outmost regions of Earth's magnetic field, a shock wave is formed as the protons and electrons are suddenly slowed to subsonic speeds. Inside this so-called *bow shock* is the magnetopause, where the pressure from the solar wind is exactly balanced by the magnetic pressure of the geomagnetic field. The region inside the magnetopause is the true domain of Earth's magnetic field.

Because of constant "blowing" by the solar wind, Earth's entire magnetosphere is swept outward, away from the sun. As shown in Figure 4-6, the magnetosphere therefore has an overall appearance of a gigantic comet. Most particles from the solar wind simply flow around the magnetosphere in the turbulent region between the bow shock and the magnetopause. But many particles pass through the magnetopause and become trapped in one of two huge doughnut-shaped radiation belts that surround our planet. The inner belt, which starts at an altitude of 2,000 kilometers above Earth's surface, contains mostly electrons. The outer belt, which is about 16,000 kilometers above Earth, contains mostly protons.

Occasionally, violent events on the sun's surface (such as a solar flare) produce an unusually large number of high-speed particles. As this "gust" in the solar wind arrives at Earth, the Van Allen belts prove unable to retain the extra particles. Delicate, shimmering auroral displays are ignited, as the surplus protons and electrons bombard Earth's upper atmosphere. In addition to aurora (often called "northern lights" or "southern lights"), radio communications are often disrupted by the influx of particles. The upper regions of Earth's atmosphere are therefore intimately associated with the outer regions of the solar atmosphere through a complex interaction of magnetic fields and charged particles.

The 1960s were marked by numerous startling discoveries about our planet. Each new satellite launched from Cape Canaveral gave some important clue or insight about the extraordinary nature and extent of Earth's magentosphere. Indeed, the picture given in Figure 4-6 would have been totally unthinkable in the 1950s. But in addition to its effects in space, Earth's magnetic field played a crucial role in providing strong evidence for a very controversial theory in-

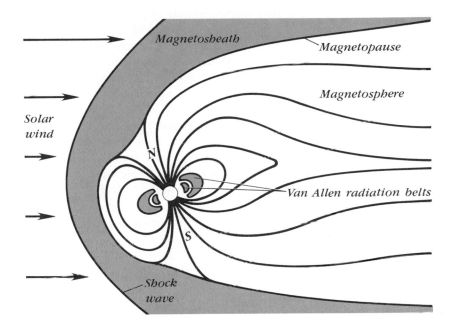

Figure 4-6   Earth's Magnetosphere
*The effects of Earth's magnetic field extend for tens of thousands of
miles out into space. A shock wave is formed where the supersonic
solar wind first encounters the outermost portion of the
magnetosphere. Some of the particles from the solar wind leak
through the magnetopause and become trapped in the Van Allen
radiation belts. (Adapted from J. A. Van Allen.)*

volving Earth's crust. In 1966, geologists and geophysicists began realizing that the continents actually move around on Earth's surface.

Lava that surges up out of a volcano is often rich in iron-bearing minerals. As the molten rock cools and solidifies, the iron atoms are permanently aligned like tiny compasses in the direction of Earth's overall magnetic field. To a small extent, the hardened lava is slightly magnetized. This "fossil magnetism" in rocks is therefore a permanent record of Earth's magnetic field at the time the rock was formed.

From studying fossil magnetism in rocks, geologists first of all discovered that Earth's magnetic field occasionally *reverses.* For some poorly understood reason, the north and south magnetic poles of our planet become interchanged. This typically occurs with a frequency of roughly once every million years. Electric currents and magnetic field lines inside Earth's liquid core become sufficiently entangled that the entire magnetic field flips over.

This remarkable discovery of reversals of the geomagnetic field had immediate practical application on the floor of the Atlantic Ocean. As shown in Figure 4-7, there is a vast chain of mountains that run the entire length of the Atlantic Ocean—from Iceland in the north to Antarctica in the south. The ocean floor consists mostly of iron-rich igneous rock. And from studying fossil magnetism in this rock, it is possible to deduce the direction of Earth's magnetic field when the rock was formed.

As data were collected from the east and west sides of the Mid-Atlantic Ridge, an astounding picture began to emerge. The ocean floor is actually spreading. As lava surges up in submarine volcanoes to form new crust, the old crust on either side of the ridge moves apart. The eastern floor of the Atlantic Ocean is moving eastward. And the western floor of the Atlantic Ocean is moving westward. The rate of separation is roughly 1 to 10 centimeters per year.

In 1910, Alfred L. Wegener proposed that long ago all the continents once fitted together to form a supercontinent called *Pangaea.* The inspiration for this idea comes from the fact that by cutting up a modern map, you can indeed fit the continents together, as shown in Figure 4-8. Then, about 200 million years ago, Pangaea began to break into two smaller supercontinents: *Laurasia* and *Gondwanaland.* Laurasia eventually fragmented into North America, Greenland,

**Figure 4-7   The Floor of the Atlantic Ocean
(Northern Portion)**
A vast mountain chain called the Mid-Atlantic Ridge runs the entire
length of the Atlantic Ocean. Molten rock rises upward in
numerous submarine volcanoes located all along the middle of the
ridge. (Courtesy of the National Geographic Society, © 1968.)

77

Figure 4-8   Fitting the Continents
Africa, Europe, Greenland, and North and South America fit together as though they were once joined. The "fit" is especially convincing if the edges of the continental shelves (rather than today's shoreline) are used. (Adapted from P. M. Hurley.)

and most of Eurasia. Gondwanaland split up into South America, Africa, Antarctica, and Australia.

Although Wegener tried to support his ideas of "continental drift" with geological and paleontological evidence (for example, that certain fossils are found in both South America and Africa as though the two landmasses were once joined), his ideas were met with vicious ridicule and derision. It was flatly inconceivable that huge continents could wander around the planet.

Today Wegener and his theory of continental drift stand totally vindicated. Actually, we now know that the continents are simply the uppermost portions of huge plates that surround our globe. Identifying the boundaries of these plates is easier than you might expect. The boundaries are the locations of violent geological activity. Volcanoes and earthquakes are found where plates are separating. In these locations, lava surges up from Earth's interior to form new crust. When plates collide, tall mountain ranges are thrust upward, accompanied by vigorous earthquakes and occasional volcanoes. In these locations, old crust plunges back down into Earth's interior. The plate boundaries are therefore found by simply plotting the epicenters of earthquakes on a map, as shown in Figure 4-9. The mechanics of plate motion is schematically diagramed in Figure 4-10.

Plate tectonics is the primary process that dominates Earth's crust. To a large extent, much of Earth's appearance is a direct result of these plate motions. For example, the plate on which India rides is vigorously colliding with the plate on which China rests. The impact pushed up a range of young mountains called the Himalayas (see Figure 4-11). By way of another example, the plates that carry Egypt and Saudi Arabia are moving apart. The resulting fissure has filled up with water and is called the Red Sea (see Figure 4-12).

Our planet is active and dynamic. Earth has been changing and evolving ever since it formed from the primordial solar nebula 4½ billion years ago. And it shows no signs of calming down. It is almost as if the planet that is teeming with life is itself alive.

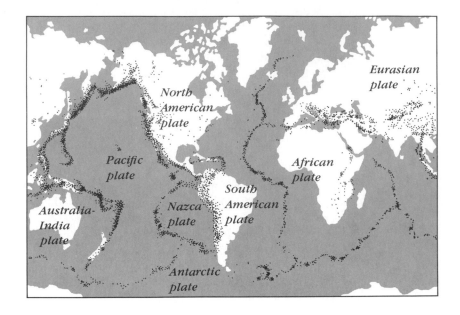

Figure 4-9   The Major Plates
*The boundaries of major plates are the scenes of violent geolog-
ical activity. Earthquakes occur when plates separate or collide.
Plate boundaries are therefore easily identified by plotting earthquakes (each dot is an epicenter) on a map.*

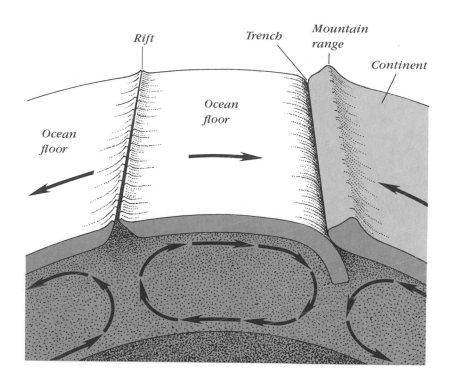

Rift

Trench

Mountain range

Continent

Ocean floor

Ocean floor

Ocean floor

Figure 4-10   The Mechanism of Plate Tectonics
*Convection currents in the plastic upper layer of the mantle
(called the asthenosphere) are responsible for pushing around rigid
plates on the crust (called the lithosphere). New crust is formed
in rifts where lava surges up between separating plates. Mountain
ranges and deep oceanic trenches occur where plates collide.*

81

Figure 4-11   The Collision of Two Plates
*The plates that carry India and China are colliding. The Himalaya
Mountains have been thrust upward as a result of this collision.
In this view, India appears at the lower left, and China is to
the right of the snow-covered mountains. (NASA.)*

Figure 4-12 The Separation of Two Plates
The plates that carry Egypt and Saudi Arabia are separating.
The Red Sea has opened up as a result of this separation. In this
view, Saudi Arabia is on the left, and Egypt (and the Nile) is on
the right. (NASA.)

5

Our Barren Moon

There are 35 moons in orbit about seven planets in the solar system. Only Mercury and Venus do not possess natural satellites. Earth is the first planet from the sun to have a moon.

A quick glance at all these moons reveals that they can be easily divided into two categories. As shown in Figure 5-1, seven of the 35 satellites could be called "giant moons." They all have diameters in the range of 3,000 to 6,000 kilometers (roughly 2,000 to 4,000 miles). They are comparable in size to the planet Mercury. In sharp contrast, all the remaining 28 satellites are quite small. These tiny moons typically measure less than 1,000 kilometers (600 miles) across. Many have diameters of only a few dozen miles. Astronomers sometimes even refer to them as "flying mountains."

It is interesting to note that six of the seven "giant moons" belong to the three largest planets: Jupiter, Saturn, and Neptune. These three planets are so huge that, by comparison, their largest satellites appear quite small. Specifically, these six moons have masses that are typically *less* than $1/1000$ of the mass of the planets that they orbit. In other words, when compared side by side with their parent planets, these six "giant moons" actually look like tiny rocks.

But the seventh "giant moon" is in orbit about Earth. Earth is one of the smaller planets in the solar system. Consequently, Earth and its moon are not terribly different in size or mass. Indeed, Earth is only 81 times as massive as its moon.

This is an unusual situation. Whereas most of the other satellites in the solar system are dwarfed by their parent planets. Earth and Moon have roughly comparable dimensions. Earth and its moon are therefore properly called a *double planet.*

The two worlds that comprise our double planet are just about as different from each other as they could possibly be. Earth has an atmosphere; the moon has none. Earth's surface is covered mostly with water; the moon is literally bone dry. Earth is teeming with life; the moon is totally barren and sterile. Earth is geologically very active: mountains rise up, volcanoes erupt, and earthquakes rumble across the landscape as tectonic plates jostle one another. But the moon has not changed appreciably for well over a billion years. Unprotected by an atmosphere, the moon's surface is bombarded by meteroids that churn up the lunar soil. And that's just about all that ever happens on the moon.

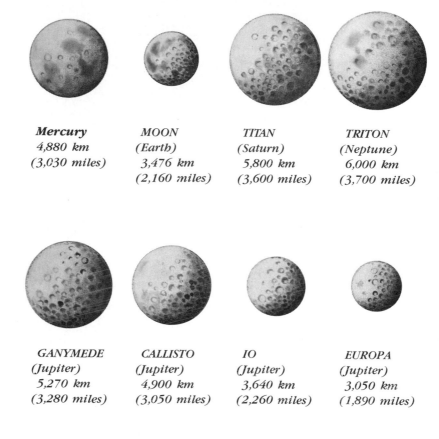

Mercury	MOON	TITAN	TRITON
4,880 km	(Earth)	(Saturn)	(Neptune)
(3,030 miles)	3,476 km	5,800 km	6,000 km
	(2,160 miles)	(3,600 miles)	(3,700 miles)

GANYMEDE	CALLISTO	IO	EUROPA
(Jupiter)	(Jupiter)	(Jupiter)	(Jupiter)
5,270 km	4,900 km	3,640 km	3,050 km
(3,280 miles)	(3,050 miles)	(2,260 miles)	(1,890 miles)

Figure 5-1 The Seven "Giant Moons"

Seven of the 35 known satellites in the solar system are very large. As shown in this scale drawing, these seven moons are comparable in size to the planet Mercury. In sharp contrast, the remaining 28 satellites are quite small.

Figure 5-2   The Double Planet
*Earth and Moon are often called a "double planet" because they
have roughly comparable sizes. In contrast, most other satellites in
the solar system are dwarfed by the planets they orbit. This
spectacular photograph, the first of its kind, was taken in 1977
from the Voyager 1 spacecraft at a distance of 11⅔ million
kilometers (7¼ million miles). (NASA.)*

Even a casual glance through an amateur telescope reveals that there are three major kinds of terrain on the moon. First of all, there are large flat planes called *seas,* or *maria.* (This unfortunate term comes from the seventeenth century, when astronomers believed that large bodies of water existed on the moon. Some imaginative observers even reported seeing ships!)

There are fourteen of these vast planes on the moon. They all have fanciful Latin names such as *Mare Tranquillitatis* (The Sea of Tranquillity), *Mare Nubium* (The Sea of Clouds), and *Mare Imbrium* (The Sea of Showers). A portion of Mare Tranquillitatis is shown in Figure 5-4. All of these planes are located on the side of the moon that perpetually faces Earth. The moon's "hidden" side has no substantial flat areas at all.

Craters are, of course, the most familiar feature on the moon. Using telescopes, earth-based astronomers have identified nearly 30,000 of these circular pits. But even with the finest earth-based telescopes, it is impossible to see lunar details smaller than a kilometer in size. As expected, close-up views during spaceflights have revealed hundreds of thousands of craters that are as small as a few feet across.

The large craters are named after prominent scientists, mathematicians, and philosophers. The crater called Copernicus is shown in Figure 5-5. Characteristic circular walls and the crater's central peak are easily seen. The moon's "hidden side" (which does not have any plains) is much more heavily cratered than the side we always see from Earth.

The largest lunar craters have diameters of 240 kilometers (150 miles). That's about the same size as Connecticut and Massachusetts combined. And the closer we look at the lunar surface, the more craters we see. There are probably over a million craters with diameters of a meter or more. Indeed, the lunar rocks themselves are sometimes covered with microscopic craters, as shown in Figure 5-6. While large craters are formed by the impact of meteoroids, these tiny craters are produced as particles of interplanetary dust strike the airless lunar surface with enormous speed.

In addition to craters and maria, there are also mountain ranges on the moon. The highest lunar peaks rise to elevations of 8 kilometers (26,000 feet) above the surrounding plains. For comparison, Mount Everest has an altitude of 29,000 feet above sea level.

Figure 5-3 The Moon (composite photograph)
*Three basic types of terrain are easily identified on the moon:
planes, craters, and mountain ranges. This photograph is a mosaic
of first-quarter and third-quarter views (so that elongated shadows
would emphasize terrain features). (Lick Observatory.)*

Figure 5-4   A Lunar Plane

Fourteen large planes, called seas *or* maria *(from the Latin* mare, *"sea") are found on the moon. All of these maria exist only on the side of the moon that perpetually faces Earth. There are no maria on the moon's "hidden side." A portion of Mare Tranquilli-tatis (The Sea of Tranquillity) is shown in this view taken by the Apollo 8 astronauts. (NASA.)*

Figure 5-5   A Lunar Crater
Nearly 30,000 lunar craters can be identified through earth-based
telescopes. Many hundreds of thousands more are viewed during
space flights. This photograph of a large crater called Copernicus
was taken by the Apollo 17 astronauts. The crater has a diameter
of 100 kilometers (60 miles). (NASA.)

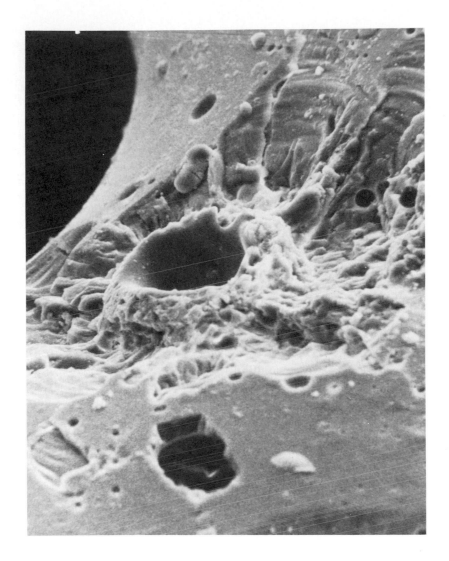

Figure 5-6   A Microscopic Crater
*Moon rocks are sometimes covered with very tiny craters. They
are typically less than half a millimeter in diameter and can be
seen clearly only with an electron microscope. These craters are
formed by the high-speed impact of meteoritic dust that momen-
tarily melts the rock's surface. (NASA.)*

Most lunar mountain ranges are named after mountain ranges here on Earth. For example, the Alps, the Caucasus Mountains, and the Carpathians are familiar sights to the veteran lunar observer. A view of Mount Hadley in the foothills of the Apennine range is shown in Figure 5-7.

Scrutiny of the lunar surface, even through an earth-based telescope, reveals many additional features and fascinating details. For example, there are valleys, crevasses, clefts, and rilles. Long, light-colored streaks, called *rays,* are found pointing away from some of the younger craters (notably Copernicus and Tycho). These rays actually consist of long, broad bands of small secondary craters that formed when debris was violently ejected from the primary crater. All of these features recount a desolate and barren world whose only scene of activity is an occasional bombardment by meteoroids, asteroids, and cometary dust.

The moon is our nearest neighbor in space. But at a distance of 384,000 kilometers (nearly ¼ million miles), it is impossible to see features smaller than a kilometer in size. Detailed examination of the lunar surface could therefore come only from space exploration.

The 1960s shall long be remembered as the decade during which humanity accomplished one of the most challenging and complex technological feats in recorded history. Tens of thousands of people labored for year after year so that by July 20, 1969, a human being could actually set foot on lunar soil.

The initial goal of lunar exploration was to get something — anything — to the moon. This was the purpose of the Ranger Program. The Ranger spacecrafts each carried six television cameras and were designed to send back pictures immediately prior to crash-landing on the moon. The first six missions ended in failure. But by 1964, all the bugs had been worked out and people around the world were treated to the first "live" TV pictures from the moon. Between July 1964 and March 1965, three Rangers sent back over 17,000 pictures of the lunar surface as they plunged toward the moon. The last of these pictures were taken from altitudes of roughly 500 meters (1,500 feet) and showed rocks and craters as small as 1 meter (3 feet) across. As demonstrated in Figure 5-8, the views were far superior to anything ever seen from Earth.

Figure 5-7   A Lunar Mountain
*There are several mountain ranges on the moon. Some lunar peaks
are as high as the Himalayas on Earth. Mount Hadley, shown in
this view by the Apollo 15 astronauts, rises to a height of 3 kilometers (roughly 10,000 feet) above the surrounding planes. (NASA.)*

Figure 5-8   A View from Ranger 9
Ranger 9, the last of the Ranger spacecrafts, impacted in the crater Alphonsus on March 24, 1965. Nearly 6,000 pictures were sent back from this mission alone. The view on the right was taken from an altitude of 415 kilometers (258 miles); the circle indicates the impact site. For comparison, the view on the left is an excellent earth-based photograph. (NASA.)

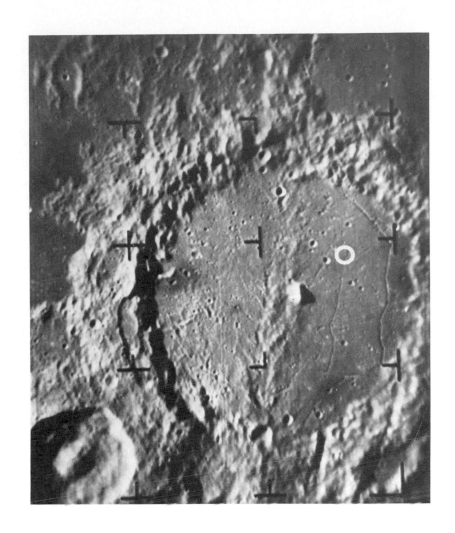

The next major effort lunar exploration consisted of two simultaneous projects: the Surveyor Program and the Orbiter Program. Between May 1966 and January 1968, five Surveyor spacecrafts successfully soft-landed on the lunar surface. Each of these three-legged spacecrafts (see Figure 5-9) was equipped with a television camera, an extendable shovel, and instruments to analyze the lunar soil. The successful landings of the Surveyors, plus the fact that they did not find any especially hazardous conditions (such as sinking into a ten-foot layer of dust as some scientists had feared) gave great encouragement to the manned space program.

While five Surveyors were gently landing on the lunar surface, five Orbiter spacecrafts were placed in orbit about the moon to perform extensive photographic surveys. All five Orbiters were successfully launched between August 1966 and August 1967 and returned a total of 1,950 excellent close-up photographs. Coverage included all of the "front" side of the moon (that is, the side that perpetually faces Earth) as well as 99½ percent of the moon's "hidden" side. It was at this time that scientists first realized that the moon's "hidden" side does not have any maria. As shown in Figure 5-10, the moon's farside is extensively cratered.

The Surveyors proved that spacecrafts could land on the lunar surface without mishap. And the Orbiter photographs were used to select a landing site for the first manned mission. The way had been paved for the Apollo Program.

A total of 24 people went to the moon (three of them made the trip twice) between December 1968 and December 1972. Twelve of these astronauts actually walked on the lunar surface. Although a wide range of geological experiments were performed, the primary achievements of the Apollo Program centers about 360 kilograms (800 pounds) of moon rocks that were brought back to Earth.

An analysis of the Apollo samples reveals that there are three types of lunar rock, each of which provides important clues about the nature and history of the moon. First of all, there is *anorthositic rock* (see Figure 5-11a). This is by far the most common kind of rock found all over the moon and is characterized by a high abundance of a mineral called plagioclase feldspar. A second important class of

Figure 5-9 Surveyor 3 and Apollo 12
Surveyor 3 successfully soft-landed in Oceanus Procellarum (The
Ocean of Storms) on February 8, 1967. Like the other four
successful Surveyors, the spacecraft sampled the lunar soil and sent
back photographs to Earth. Surveyor 3 was visited 2½ years later by
the Apollo 12 astronauts. (NASA.)

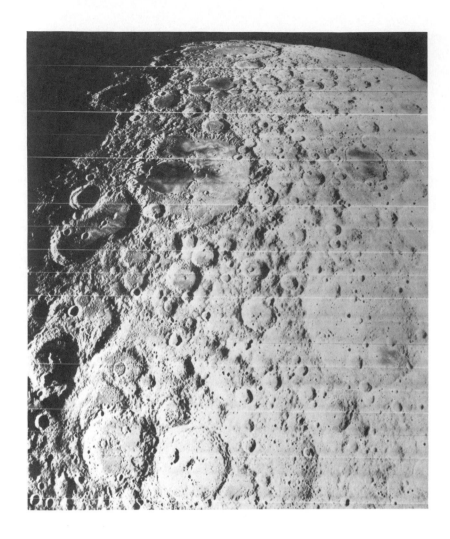

Figure 5-10   The Lunar Farside from Orbiter 2
Five Orbiter spacecrafts were successful in obtaining nearly complete photographic coverage of the entire lunar surface. Objects as small as 10 meters (30 feet) across could be seen. These missions revealed that the moon's farside is covered with nothing but craters. This photograph of the heavily cratered farside was taken from an altitude of 900 miles above the lunar surface. (NASA.)

lunar rock is called *"KREEP" norite* because of its high content of potassium (K), rare earth elements (REE) and phosphorus (P). KREEP norite (see Figure 5-11b) is typically found in the light-colored mountainous regions of the moon. In contrast, the dark-colored "seas," or maria, are covered with *mare basalt* (see Figure 5-11c).

Anorthositic rock is the most abundant and most ancient type of rock found on the moon. Data from seismometers (left by astronauts on the lunar surface) as well as remote geochemical analysis from lunar orbit reveal that the moon's crust down to a depth of 60 kilometers (40 miles) is predominantly anorthositic. Anorthosite has the highest melting temperature of the three basic types of lunar rocks. Anorthositic rock was therefore the first rock to solidify as the moon's primordial molten surface began to cool.

Prior to the Apollo Program, there were three competing theories about the origin of the moon. Some scientists felt that the moon might simply have been captured by Earth. Others believed that primordial Earth may have fissioned or broken into two pieces (the Pacific Ocean was presumably the "hole" left behind as the moon spun outward from Earth). But analysis of lunar rocks seem to favor the idea that the moon formed from the accretion of tiny rocks that orbited Earth 4½ billion years ago. As shown in Figure 5-12, the gravitational accretion of particles about Earth was a small-scale version of the accretion processes in the primordial solar nebula that gave birth to the planets.

The moon's birth must have been a violent and dramatic event that may have taken only a few thousand years. As millions upon millions of earth-orbiting rocks plunged into the ever-growing primordial moon, the entire lunar surface must have been a molten sea of white-hot lava. But once most of the rocks had been swept up, the lunar surface could begin to cool and solidify. It was at this time, 4½ billion years ago, that the moon's anorthositic crust started to form.

Both KREEP norite and mare basalt have lower melting temperatures than anorthositic rock. The existence of these two younger types of lunar material must therefore signify important events later in the moon's history.

101

a

Figure 5-11a   Anorthosite
Anorthositic rock is by far the most abundant and the most ancient type of rock found on the moon. The entire lunar crust down to a depth of 60 kilometers (40 miles) is predominantly anorthosite. (National Space Science Data Center.)

Figure 5-11b   KREEP Norite
KREEP norite is a type of lunar rock named for its high abundance of potassium (K), rare earth elements (REE), and phosphorus (P). KREEP norite is typically found in the mountainous regions of the moon. (National Space Science Data Center.)

Figure 5-11c   Mare Basalt
The vast, dark-colored planes that dominate the moon's nearside are covered with mare basalt. Mare basalts are among the youngest rocks found on the moon. (National Space Science Data Center.)

b

c

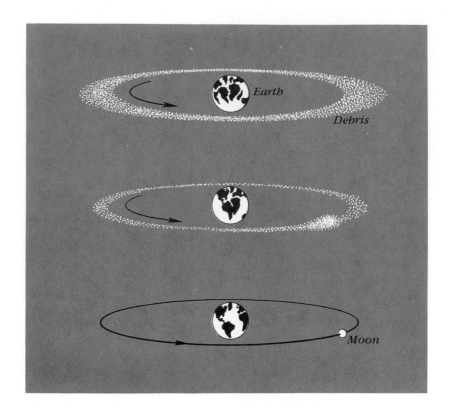

Figure 5-12   The Creation of the Moon
Rocks of all shapes and sizes were gravitationally captured by primordial Earth as our newly formed planet swept through the debris of the solar nebula. Gradually, all these Earth-orbiting particles collided and accreted to form the moon. (Adapted from M. Zeilik.)

KREEP norite is characterized by a high abundance of elements whose atoms tend to be rather large. Because of their exceptionally large size, these atoms are not easily incorporated into the crystals that make up a specimen of anorthosite. In other words, if anorthositic rock is heated and subjected to *partial melting,* these large-atomed substances would be preferentially "sweated out." It is therefore reasonable to suppose that KREEP norite was formed by the partial melting of anorthositic rock.

KREEP norite is found in the mountainous regions of the moon. Scientists still do not know how the lunar highlands were formed. But the same violent processes that thrust up the lunar mountain ranges could have also partially melted the recently formed anorthositic crust some 4 billion years ago. This would explain the abundance of KREEP norite in mountain ranges like those that border Mare Imbrium and Oceanus Procellarum.

It is obvious that the moon has been pelted by enormous numbers of meteoroids over the ages. That's why there are so many craters. But the maria are by far the largest impact features on the lunar surface. Perhaps 3½ to 4 billion years ago, at least a dozen asteroid-sized objects violently collided with the moon. These planet-shattering impacts excavated enormous craters that broke through to the young moon's liquid interior. Lava welled up from the lunar depths, and in only a few hundred thousand years, the colossal excavations were filled in. The dark, flat maria were formed as flood after flood of molten rock healed the wounds inflicted by asteroids. This was the origin of mare basalt — the youngest of the major types of lunar rock.

The lunar crust must be thinner on the moon's earthside than on the farside (see Figure 5-13). Major impacts by planetesimals on the lunar farside did not succeed in penetrating the moon's crust. Consequently, there were no extensive lava floodings and thus no mare formation on the side of the moon that is hidden from earthview.

Not much has happened on the moon over the past 3 billion years. Meteoroids still rain down on the lunar surface (although in far fewer numbers than in eons past). This constant fine-grained bom-

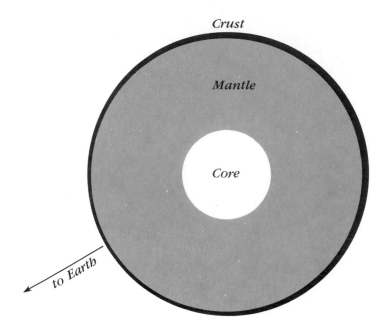

Figure 5-13   The Structure of the Moon's Interior
*Like Earth, the moon has a core, a mantle, and a crust. Notice
that the crust on the lunar earthside is thinner than on the farside.
This explains why there are no maria on the moon's farside.
Meteoroids could not puncture the thick farside crust and expose
the once-molten interior. (Adapted from J.A. Wood.)*

bardment gradually churns up the lunar soil, or *regolith,* as it is properly called.* A big meteoroid has not hit the moon since the kilometer-sized rocks that formed the rayed craters called Copernicus and Tycho.

Lunar exploration has revealed a barren and sterile world vastly different from Earth. But because of all its vitality, Earth's early history has been almost totally obliterated by the relentless action of wind, rain, and snow. In sharp contrast, some of the most ancient events in the solar system are preserved for all eternity on the airless, lifeless surface of our planet's nearest neighbor.

---

*The word "soil" connotes a substance that contains decayed biological material. The term "regolith" simply refers to a ground up layer of rock.

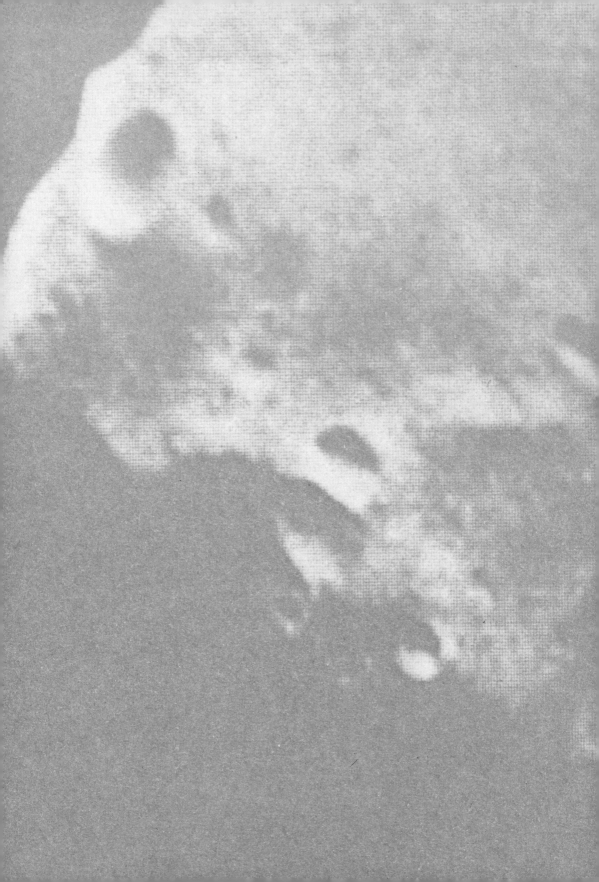

6

The Martian Invasions

Life is a common phenomenon on the third planet from the sun. The exceptional ability of one long double-helixed molecule to manufacture exact replicas of itself has populated our planet with a mind-boggling plethora of creatures. It all supposedly began 3 to 4 billion years ago in a warm, murkey primal sea. Bits and pieces of organic molecules, amino acids, and nucleotides happened to combine to form that first crucial strand of deoxyribonucleic acid.

But why should Earth be the only place where this extraordinary biochemical synthesis has occurred? Perhaps similar life-giving molecules are created on any planet where appropriate conditions exist. Or alternatively, the appearance of the first living cells on Earth might be far more improbable than we could ever imagine. This would mean that Earth is a totally unique experiment in an otherwise lifeless universe.

In the mid-1970s, scientists began to reexamine these fascinating speculations in the light of experimentation and observation. The most profound of all questions — "Are we alone?" — has finally come of age. Humanity has taken the first bold steps to explore new worlds in search of extraterrestrial life.

Traditionally, Mars has been the place to look for extraterrestrial life. Mars is the only planet whose solid surface can be easily viewed through earth-based telescopes. These views reveal that the tiny reddish world has some interesting and suggestive earthlike qualities.

First of all, Mars rotates about its axis in 24⅝ hours. This means that a day on Mars (properly called a "sol") is only 41 minutes longer than a day on Earth. In addition, Mars's axis of rotation is tilted by 24° from the perpendicular direction to its orbit (in the case of Earth, the angle happens to be 23½°). This means that the Martian surface experiences seasons similar to the seasons of Earth. Of course, the seasons last nearly twice as long on Mars because the planet takes 687 days (1⅞ earth-years) to circle the sun.

These superficial similarities between Earth and Mars are, of course, purely coincidental. But ever since the late 1700s, astronomers have observed seasonal changes on the Martian surface that are intriguing to the eye and stimulating to the imagination.

Like Earth, Mars has two polar caps. These whitish areas (see Figures 6-1 and 6-10) are large during the Martian winter. But they

Plate 1   The Chryse Plains
*This view looks toward the southeast as seen from Viking
Lander 1. (NASA; JPL.)*

Plate 2   The Utopia Plains
*This view looks toward the northeast as seen from Viking
Lander 2. (NASA; JPL.)*

Plate 3   Summertime at Utopia
Portions of the Viking lander are seen in this southward-facing view. (NASA; JPL.)

Plate 4   Wintertime at Utopia
Patches of frost are seen in this late-winter view from Viking Lander 2. (NASA; JPL.)

Plate 5 Jupiter from Pioneer 10
*This photograph was taken at a distance of 2½ million
kilometers (1½ million miles) from the planet. The Great Red Spot
and the shadow of Io are clearly seen. (NASA.)*

Plate 6 Jupiter's North Pole from Pioneer 11
*This photograph was taken at a distance of 1⅓ million kilometers
(¾ million miles) from the planet. Notice that the belts and
zones break up into random eddies in the polar regions. (NASA.)*

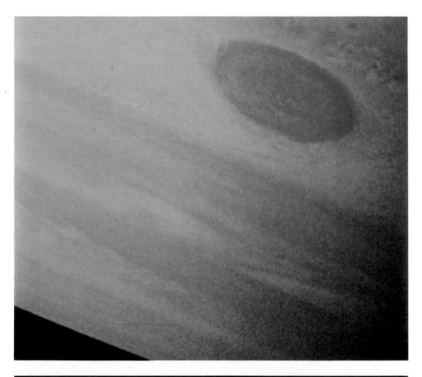

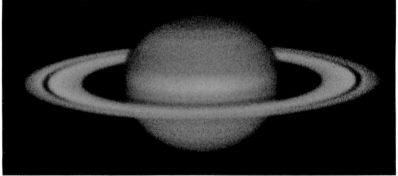

Plate 7 The Great Red Spot
Three Earths could fit side by side across the Great Red Spot. This familiar feature is actually a huge, high pressure hurricane that has persisted for at least three centuries. (NASA.)

Plate 8 Saturn
Faint suggestions of belts and zones are seen in this earth-based photograph. Cassini's division dominates the appearance of the rings. (NASA.)

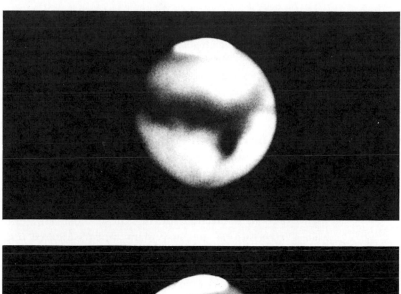

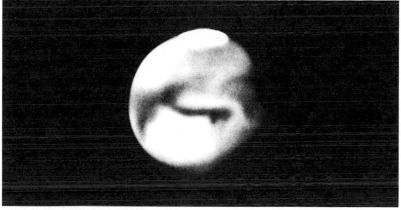

Figure 6-1 Mars (Two Earth-Based Views)
*Telescopic examination reveals that Mars has many earthlike
qualities. There are white polar caps that vary in size according to
the seasons. And there are dark markings that appear greenish
against the red Martian surface. These markings seem to vary in
brightness with the seasons on Mars in a way that suggests the
possibility of vegetation. (Lick Observatory.)*

shrink and nearly fade from view with the coming of the Martian summer. In addition, there are dark markings on Mars that also exhibit seasonal variations. These markings seem to be more pronounced during the Martian spring and summer. But the features become pale and wan as the Martian fall turns to winter. In contrast to the reddish hue of the soil, these dark markings appear to have a distinct greenish color. This obviously suggests vegetation, and consequently, ever since the nineteenth century, the certainty of life on Mars was almost unquestionable.

There are good times to observe Mars through earth-based telescopes. And there are bad times. As diagramed in Figure 6-2, Earth catches up with Mars every 780 days. But Mars's orbit about the sun is significantly more elliptical than Earth's. This means that the Earth–Mars distance during these close encounters (properly called "oppositions" because Mars appears located opposite the sun in the sky) can be as small as 56 million kilometers (35 million miles). Or it can be as large as 101 million kilometers (63 million miles). The best earth-based views of Mars are obtained only during "favorable opposition," when the distance to the tiny red planet is as short as possible.

During one of these favorable oppositions in 1887, the Italian astronomer Giovanni Schiaparelli reported seeing straight lines crisscrossing the Martian surface. Italian astronomers called these features "canali," which was incorrectly translated into English as "canals." Canals, like those in Panama and the Suez, are waterways constructed by intelligent creatures. Since canals had been seen on Mars, it logically followed that intelligent creatures inhabit the planet. And soon certain enthusiastic astronomers, especially Percival Lowell in Arizona, began discovering dozens of canals all over Mars. The obvious explanation for this planetwide system of canals involved the transporting of water from the melting polar caps to vegetation in the equatorial regions! Mars surely is an arid and dying planet whose inhabitants must perform technological miracles just to water their crops. Surely they would be envious of our Earth. If they could build all those canals, they would have no trouble with interplanetary space flight. We would be defenseless against the onslaught. They would crush us like ants. The final terrifying invasion would soon begin.

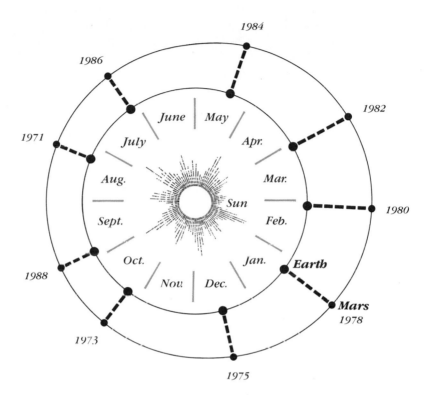

Figure 6-2 The Orbits of Earth and Mars
*Earth catches up with Mars every 780 days (nearly 2¹⁄₇ years). These
close encounters (properly called "oppositions") are the best times
to observe the red planet. But because Mar's orbit is noticeably
elliptical, some oppositions afford better views than others. The
next "favorable opposition" will occur in September 1988.*

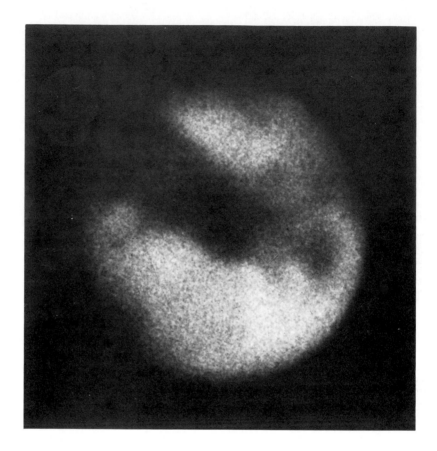

Figure 6-3 Martian Fantasies

During moments of "good seeing," when the air is exceptionally calm and clear, some earth-based observers have claimed to see networks of lines crisscrossing the Martian surface. Some imaginative people have suggested that these lines (called "canals") are irrigation ditches constructed by intelligent creatures. This drawing and the corresponding photograph were made at the same time during the opposition of 1926. (Lick Observatory.)

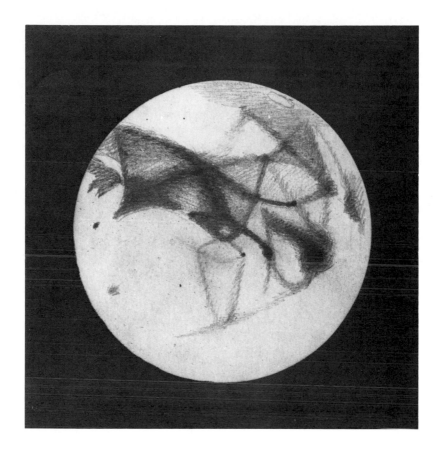

And it did. In books, in magazines, and in an endless stream of Grade B movies. We were blasted and roasted and eaten and beaten as wave upon wave of horrible creatures overran our planet. Unfortunately, Schiaparelli's "canali" do not exist. They never did. Like the Martians themselves, they were entirely figments of imaginations.

The shattering blow to all this fantasy came in 1965 when the spacecraft called Mariner 4 flew within 10,000 kilometers of the Martian surface. Twenty-two close-up pictures were sent back to Earth. To everyone's surprise, these views revealed a cratered landscape that was noticeably more reminiscent of our moon than Earth. But unlike lunar craters, these Martian craters are very flat-bottomed. Apparently, the craters have been partly filled in during occasional dust storms that sweep across the planet.

Although Mariner 4 photographed only 1 percent of the Martian surface, the entire mission was so successful that scientists promptly vowed to return to the red planet. And they did. In February and March of 1969, two more spacecrafts coasted over the mysterious Martian landscape. Together, Mariners 6 and 7 sent back 200 photographs that showed that Mars has its own unique character. In addition to the flat-bottomed craters (see Figure 6-4), the photographs also revealed vast featureless planes as well as hilly jumbled terrain.

The three historic Mariner missions to Mars in the 1960s discovered a hostile, arid environment. The air pressure in the thin carbon dioxide atmosphere is less than $1/100$ of the air pressure at sea level here on Earth. And at best, the temperature at noon on the Martian equator creeps up to 10°C (50°F). But at the polar caps during winter, temperatures plunge to −120°C (−185°F). In these brutally cold locations, carbon dioxide in the atmosphere actually freezes and falls to the ground like snow. The expansive polar caps that can be seen so easily through earth-based telescopes are made almost entirely of dry ice.

While the visions of imaginative science fiction writers had been permanently shattered, astronomers found themselves with mysteries, enigmas, and a host of unanswered questions. Unfortunately, Mariners 4, 6, and 7 had merely skimmed past Mars for a brief

Figure 6-4   Martian Craters
Numerous flat-bottomed craters are seen in this mosaic of four views taken from Mariner 6 in 1969. The area shown in these photographs measures roughly 3,500 by 700 kilometers (2,200 by 400 miles) and is located just south of the Martian equator. (NASA.)

glance, never to return. It obviously would be far better to go into orbit about the planet for prolonged observation.

The entire Martian surface was embroiled in a planetwide dust storm when Mariner 9 arrived. Meterologists had a field day, but everyone else had to wait until early 1972 for the dust to settle. And as the atmosphere cleared, Mariner 9 began sending back incredible photographs of terrain that no one in his or her wildest dreams had ever imagined.

Olympus Mons was first to poke its summit through the dust storm. Olympus Mons is a volcano. It is absolutely enormous. There is simply nothing on Earth that could compare with the colossal size of Olympus Mons. It is almost three times as high as Mount Everest. The volcano's base (see Figure 6-6) terminates with a sharp cliff roughly 500 kilometers (300 miles) in diameter. If Olympus Mons were located in California, it would just barely fit between Los Angeles and San Francisco.

Olympus Mons is one of four large volcanoes that are grouped together in the so-called Tharsis Region, just north of the Martian equator. A second, less prominent grouping of volcanoes is located about 5,000 kilometers away in the so-called Elysium Region. For some unknown reason, all volcanic activity on Mars seems to be restricted to the planet's northern hemisphere. In contrast, the majority of the flat-bottomed craters are located south of the equator.

In addition to gigantic volcanoes, photographs from Mariner 9 soon revealed a vast canyon. Once again, there is nothing on Earth that could begin to compare with the Martian canyonlands. There is a canyon on Mars, called Valles Marineris, that is 4,000 kilometers long (2,500 miles), up to 200 kilometers wide (120 miles), and plunging to depths of 6 kilometers (20,000 feet) below its rim. A small portion of this chasm is shown in Figure 6-7. For comparison, the Grand Canyon is only 150 kilometers long and at most 2 kilometers (less than 7,000 feet) deep. If Valles Marineris were located on Earth, it would reach all the way from New York to Los Angeles!

The volcanic and canyon features discovered from Mariner 9 strongly suggest that Mars is a planet where tectonic plate motions never got started. Perhaps Valles Marineris is a huge rift where two

118

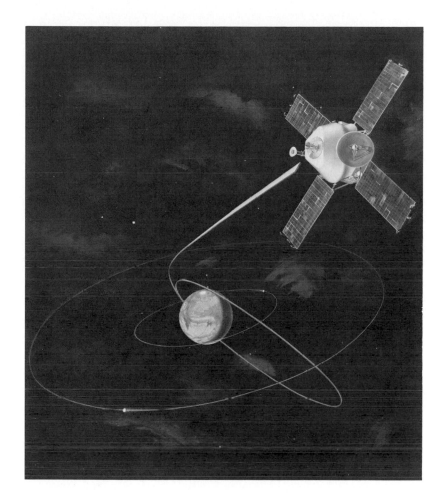

Figure 6-5   The Orbit of Mariner 9
*Mariner 9 became the third moon of Mars on November 14, 1971.
From its highly elliptical 12-hour orbit, Mariner 9 managed to
photograph the entire planet. A total of 7,329 photographs were
transmitted back to Earth. (NASA.)*

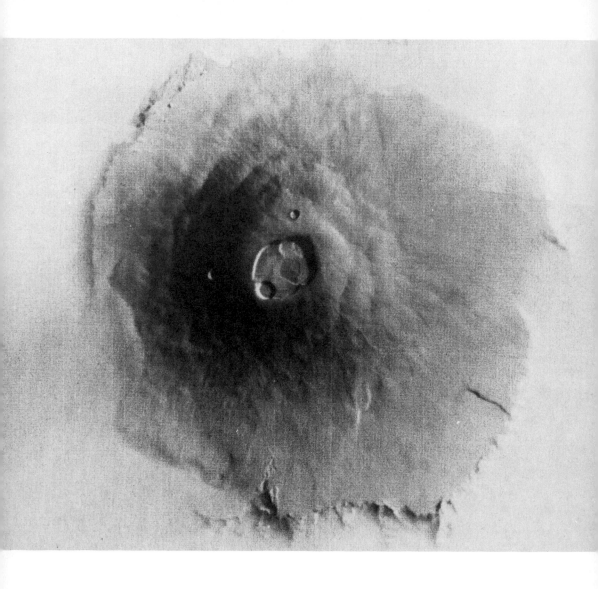

Figure 6-6 Olympus Mons

Olympus Mons is the largest volcano on Mars. The summit is at an elevation of 24 kilometers (78,000 feet) above the surrounding planes. That's nearly three times a high as Mount Everest. Olympus Mons is so broad at the base that it could cover the entire state of Missouri. (NASA.)

120

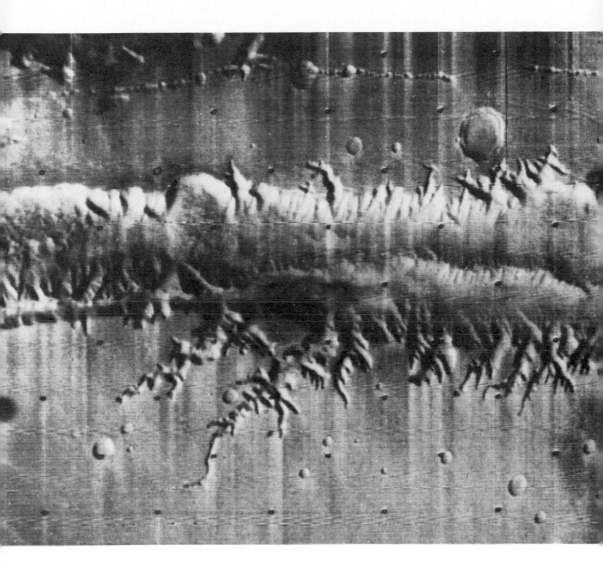

Figure 6-7   Valles Marineris
Only a small segment of Mars's Grand Canyon is shown in this photograph. The entire canyon is 4,000 kilometers (2,500 miles) long and runs nearly parallel to the Martian equator. The area covered in this view measures 500 by 400 kilometers (300 by 240 miles). (NASA.)

plates began to move apart a couple billion years ago. But because of Mars's small size, the planet cooled rather rapidly and its crust soon became very thick — much thicker than Earth's crust. So the would-be plates simply bogged down. With no plate motions, hot spots deep in the Martian mantle kept squirting lava to the surface in the same place year after year. Thus, instead of producing a long chain of small volcanoes (like the Hawaiian-Emperor chain beneath the Pacific Ocean), a few very large volcanoes were created.

Almost 400 years ago, the Renaissance astronomer Johannes Kepler argued that Mars must have two moons. After all, Earth has one moon and Galileo had just discovered four moons orbiting Jupiter. The idea that Mars possessed two moons appealed to Kepler's sense of order and harmony in the universe.

In spite of careful searches over the years, the Martian satellites evaded astronomers until the favorable opposition of 1877. In that year, Asaph Hall at the U.S. Naval Observatory managed to sight two faint specks of light very near the red planet. He named the moons Phobos and Deimos — meaning "fear" and "panic" — appropriate mythical companions of the god of war. Because of their faintness, it was immediately clear that both moons are very small.

Both Phobos and Deimos travel around Mars in nearly circular orbits not far above the planet's equator. In fact, their orbits are so close to the planet that neither moon can be seen from the Martian polar regions.

Phobos, the larger and nearer of the two moons, races around the planet in only 7½ hours at a distance of 5,000 kilometers (3,000 miles) above the Martian surface. Since this orbital period is much shorter than the length of the Martian day, Phobos appears to rise in the west and gallop across the sky in only 5 hours, as viewed by an observer near the Martian equator. During this time, Phobos would appear several times as bright as Venus does from Earth.

Deimos, which is farther from Mars and somewhat smaller than Phobos, would appear in the Martian sky to have a brightness roughly the same as Venus does from Earth. But Deimos's orbit, 20,000 kilometers (12,500 miles) above the Martian surface, is almost at the right distance for a synchronous orbit — an orbit in which an object appears to hover above a single location on a planet's equator. Consequently, as seen from the Martian surface, Deimos takes almost three full days to creep laboriously from one horizon to the other.

During the Mariner 9 mission, astronomers were treated to the first close-up views of the Martian moons. Twenty-seven pictures of Phobos and nine of Deimos were sent back to Earth. One of the best photographs is shown in Figure 6-8. Four years later, the Viking orbiters managed to glimpse surface details on these moons with unprecedented clarity, as shown in Figures 6-9 and 9-2. Phobos and Deimos were revealed to be jagged, heavily cratered rocks. For the first time, scientists could directly measure the sizes of these football-shaped satellites. Phobos is roughly 27 by 21 by 19 kilometers ($17 \times 13 \times 12$ miles). Deimos is slightly smaller, with dimensions of roughly 15 by 12 by 11 kilometers ($9 \times 7 \times 7$ miles). These observations further revealed that both Phobos and Deimos rotate synchronously about Mars. In other words, each satellite keeps the same side facing the red planet, just as our moon keeps the same side exposed to earth-view.

No one knows where the Martian moons came from. Some scientists feel that Phobos and Deimos might have formed by the gravitational accretion of tiny rocks that orbited Mars in the distant past. In that case, the creation of these two satellites would have mimicked the formation of our own moon, but on a much smaller scale. If this is correct, then the Martian moons are probably composed of basaltic rock like our own satellite.

A second possibility is that Phobos and Deimos are captured asteroids. As we shall see in Chapter 9, Mars is quite near the asteroid belt, where thousands of large rocks orbit the sun. It is possible that two of these asteroids happened to wander close enough to Mars to become permanently trapped by the planet's gravitational field. If this is correct, then the Martian moons are probably composed of chondritic material, like many meteorites. But basaltic rock is slightly denser than chondritic rock. Accurate measurements of the densities of Phobos and Deimos (that is, dividing their masses by their volumes as determined by the Viking orbiters) might settle the controversy.

By far, the most controversial discovery made by Mariner 9 centered about features that look just like dried-up river beds. This was very puzzling, because Mars is incredibly arid. Unlike our own

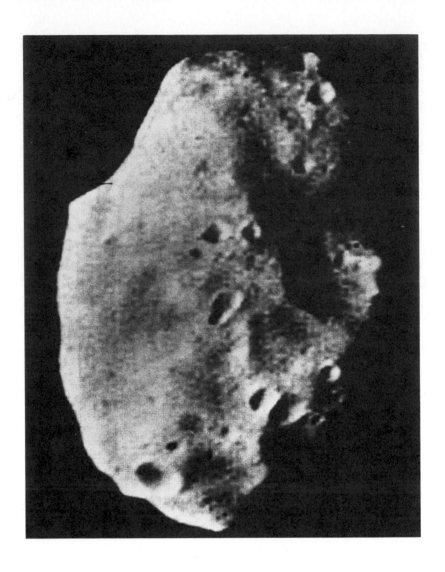

Figure 6-8   Phobos

More than a dozen craters are seen in this view of Mars's larger moon. The moon measures only 27 kilometers (17 miles) in length, and the presence of craters suggests that the satellite is very old. Mariner 9 was 5,540 kilometers (3,440 miles) from Phobos when this picture was taken. (NASA.)

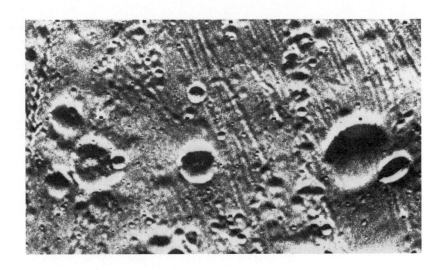

Figure 6-9   Craters on Phobos
Features as small as 40 meters (130 feet) in size can be recognized in this extraordinary close-up photograph of Phobos. Nearly a fifth of the tiny moon's surface is seen in this view, which was taken from a distance of only 880 kilometers (550 miles) by one of the Viking orbiters. (NASA.)

sopping wet planet, there is not one single drop of water on the Martian surface. And if all the water vapor in the Martian atmosphere were condensed out, it would hardly be enough to fill one good-sized lake. Mars is literally bone dry. So how could there be water erosion without any water?

Important clues to the "mystery of the missing water" came with the arrivals of Viking 1 and Viking 2 at Mars in the summer of 1976. Both spacecrafts spent many weeks taking close-up high-resolution photographs of the planet in search of suitable landing sites. These reconnaissances disclosed many more signs of water erosion than the few dried river beds seen from Mariner 9. As shown in Figure 6-10, these features clearly indicate flash flooding. It seems that on rare occasions, vast quantities of water suddenly and briefly swept across sections of the Martian surface.

125

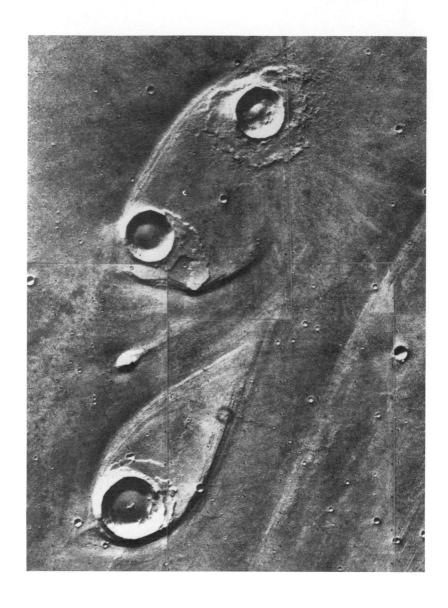

Figure 6-10 Water Erosion on Mars
*Millions of years ago, a flash flood swept over this region of
the Chryse planes. Except for a few high-rimmed craters, nearly
everything was obliterated by the onrushing water. (NASA.)*

The cameras onboard Viking 1 and Viking 2 also photographed several impressive landslides and regions where the Martian surface have collapsed. One of these dramatic landslides is shown in Figure 6-11. The appearance of these landslides and collapsed areas strongly suggests that the Martian soil had once been in a semiliquid state, like mud. And in several cases, water erosion features are seen emanating directly from the collapsed terrain.

Like the arctic tundra here on Earth, it seems reasonable to suppose that there may be substantial deposits of subsurface ice and permafrost on Mars. On occasion, volcanic activity melts the ice or permafrost. This activity undermines the Martian surface, and tons of overlaying rock and soil collapse downward into the pool of water. The water therefore vigorously bursts to the surface, and a flash flood results.

Water cannot exist for long on the Martian surface. The atmospheric pressure is so low that any water rapidly boils away. But at the poles, the temperature is so low that water-ice can survive even through the summer seasons. Mariner 9 had shown that the polar caps consist of dry ice that blankets the ground when carbon dioxide freezes out of the atmosphere. Instruments onboard Vikings 1 and 2 established the existence of permanent water-ice polar caps beneath the veneer of carbon dioxide snow.

The primary purpose of the Viking missions was to land spacecrafts on the Martian surface, analyze the soil, and search for life forms. This was accomplished by two "landers" that separated from the "orbiters" and parachuted down to the ground. Each lander (see Figure 6-13) was equipped with two television cameras, a mechanical arm and scoop to deliver soil to experiments inside the spacecraft, and several meteorological instruments.

As shown in Figure 6-14 and Plates 1 through 4, the rock-strewn Martian landscape is totally barren. In looking at these photographs, we are struck by the complete absence of any signs of life. This sterile appearance of the Martian environment was eventually confirmed by the biology experiments onboard the landers — but not without considerable debate and controversy.

The Viking landers each carried three biological experiments that were designed to detect microorganisms in the Martian soil. All

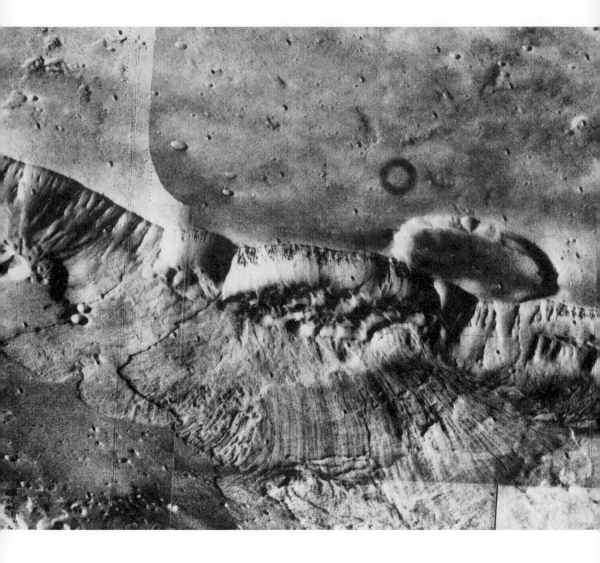

Figure 6-11   A Landslide
A 70-kilometer (40-mile) long section of the south wall of Valles
Marineris is shown in this remarkable photograph from Viking 1.
The canyon wall obviously collapsed in a huge landslide that
flowed down and across the canyon floor. (NASA.)

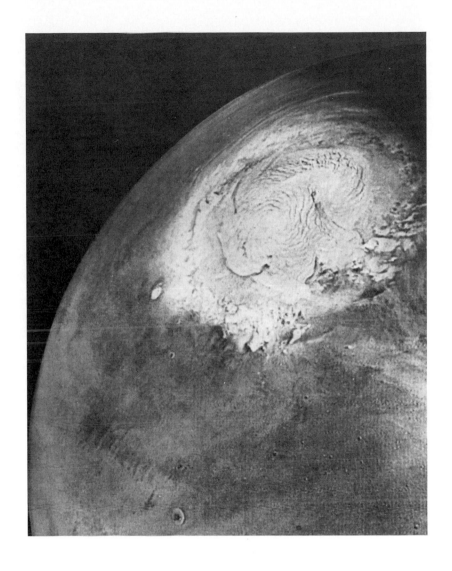

Figure 6-12   The North Polar Cap
Most of the Martian polar caps that cover thousands of square miles during the winter seasons consist of a thin layer of carbon dioxide snow. This powdered dry ice completely evaporates during the Martian summer. A small, permanent water-ice polar cap always survives through the warm seasons. This Mariner 9 photograph was taken during the Martian springtime. (NASA.)

Figure 6-13   A Viking Lander
*Two of these spacecrafts successfully soft-landed on Mars during
the summer of 1976. Viking 1 landed in the Chryse Planes on
July 20, 1976. And Viking 2 set down in the Utopia Planes on
September 3, 1976. Both landing sites are in the northern hemi-
sphere; they are separated by 7,500 kilometers (4,600 miles).
(NASA.)*

three experiments were built around the premise that *living organisms alter their environment.* They eat, they breathe, and they give off waste products. All three experiments therefore involved placing a sample of Martian soil into a closed container (perhaps along with some water or nutrient) and then looking for some changes inside the container.

The "labeled release experiment" was designed to detect *metabolism.* A pinch of Martian soil was placed in a container. The soil was then moistened with a nutrient (affectionately called "chicken soup") that contained radioactive carbon. If there are any organisms in the soil, they should eat the food and emit gases containing the telltale radioactive carbon. Scientists were amazed to find that large quantities of radioactive gas were released from unsterilized soil. It looked as though they were on to something!

The "pyrolytic release experiment" was designed to detect *photosynthesis.* Again, a small sample of Martian soil was placed in a container. The sealed container was then filled with a Martian-like atmosphere, except that the carbon dioxide contained radioactive carbon. And finally, the container was illuminated with artificial sunlight. If any photosynthesis occurs, radioactive carbon in the air will become incorporated into the living organisms. To test for this, the gases were first flushed from the container and the remaining soil was analyzed for radioactive carbon content. In some cases, the results were positive. And, confusingly, at other times the results were negative.

The "gas exchange experiment" was designed to detect *respiration.* A small sample of soil was placed in a sealed container along with a controlled amount of gas. The soil was then moistened with nutrient (more "chicken soup"), and the gases in the container were periodically monitored. If living organisms are present in the soil, they should inhale and exhale gases, thereby changing the composition of the atmosphere in the container.

The gas exchange experiment was the first to dampen any enthusiasm about the possibility of having discovered Martian life. This experiment showed that as soon as the Martian soil is moistened, substantial amounts of carbon dioxide and oxygen are rapidly given off. Although this might sound like a positive biological result,

Figure 6-14   The First Photographs
*The first photographs from Viking 1 (upper) and Viking 2
(lower) were transmitted to Earth immediately after the successful
landings. The rocks in both of these views have sizes in the range
of 10 to 20 centimeters (4 to 8 inches). (NASA.)*

the *rate* at which these gases are produced is exactly what you would expect from a purely chemical reaction. In other words, the Martian soil must contain some sort of peroxides or superoxides that fizz and bubble when moistened with water. The reaction is similar to dropping an Alka-Seltzer tablet into a glass of water. The apparently positive results from the labeled release experiment can be explained in this same way.

Although the Viking missions have failed to detect any signs of life, they have established that the Martian soil is unusually chemically active. Unlike Earth, Mars does not have a protective layer of ozone high in its atmosphere to filter out the sun's ultraviolet radiation. This energetic radiation, which is deadly to terrestrial organisms, may be responsible for chemically activating the soil so that it effervesces when moistened with water.

But maybe we have not discovered life on Mars simply because we performed the *wrong* experiments. Perhaps Martian organisms do not function by metabolism, photosynthesis, or respiration in a way at all analogous to earth-based creatures. Or maybe we just looked in the wrong places. There is an enormous amount of totally unexplored territory that may be radically different from the two Viking sites.

Scientists are anxiously making plans for a return trip to Mars in 1984. This time, the lander (properly called a "rover") will be equipped with wheels. The spacecraft will probably cost $800 million. For comparison, in 1976 (the year of the Viking missions) Americans spent $800 million on chewing gum.

We have only begun to scratch the surface of this intriguing red world. And indeed, there may be no life on Mars after all. Mars could easily be just as sterile and barren as our moon. Like household hydrogen peroxide that you pour on a cut or a wound, the chemically active Martian soil may have ensured that the entire planet is perpetually disinfected. And we would have to focus our search for extraterrestrial life elsewhere — perhaps in the turbulent reddish-brown clouds in Jupiter's atmosphere.

7

Lord of
the Planets

Perhaps more than any other single factor, the temperatures throughout the primordial solar nebula dictated the final fate and form of the planets that now orbit our star. In the warm, inner regions of this ancient nebula, dust grains consisted primarily of metals, oxides, and silicates. The temperature was simply too high to allow substantial condensation of volatile substances like water, methane, and ammonia. The four planets that formed close to the primal sun were therefore destined to be composed almost entirely of rocky material. Although the solar nebula consisted mostly of hydrogen and helium, none of these lightweight gases could be retained by the tiny inner planets. Their surface gravities were too low and their surface temperatures were too high. And the hydrogen and helium easily drifted out into interplanetary space.

But it has always been cold at a distance of half a billion miles from the sun. At the orbit of Jupiter, five times the distance from the sun that Earth is, sunlight is dimmed to only $1/27$ of the brightness we observe from our planet. Even in the most ancient times, temperatures were low, and consequently the primordial dust grains were covered with thick frosty coatings of water-ice and frozen methane and ammonia. These volatile substances were therefore destined to become important constituents of the remote worlds that circle the sun.

At half a billion miles from the protosun, the coalescence of ice-covered primordial dust grains must have proceeded at a furious rate, far more efficiently than in the inner regions of the solar nebula. A large planetary core, the primal seed around which Jupiter would grow, was soon created. But then, because of the low temperature and the strong surface gravity of this massive planetary core, substantial quantities of hydrogen and helium were gravitationally attracted toward the young protoplanet. Detailed calculations of this process by Fausto Perri and Alastair Cameron reveal that the accretion of hydrogen and helium became hydrodynamically unstable. In other words, as soon as the planetary core had grown to a certain critical size, all of the surrounding hydrogen and helium rapidly collapsed onto the protoplanet. The final result was the largest planet in the solar system—a planet composed of 82 percent hydrogen, 17 percent helium, and 1 percent everything else—essentially the same

Lord of the Planets

abundance of materials that were present in the primordial solar nebula. This same overall scenario is thought to account also for Saturn's large size (and perhaps Uranus and Neptune, but to a much lesser extent).

Jupiter is huge. Its diameter is 11¼ times as large as the diameter of Earth. This means that 1,400 Earths could fit inside Jupiter if it were hollow.

Jupiter is massive. Its mass is 318 times as great as the mass of Earth. Jupiter is 2½ times more massive than *all* the other planets, satellites, asteroids, meteoroids, and comets put together.

Jupiter is rapidly rotating. In fact, Jupiter spins faster than any other planet in the solar system. A "day" on Jupiter lasts only 9 hours and 55 minutes. Actually, the planet does not rotate like a rigid object. The equatorial regions rotate a little faster (9 hours and 50 minutes) than the temperate and polar regions. This is possible, of course, because the planet is not made out of solid material.

Jupiter's rapid rotation is directly responsible for many of the planet's intriguing properties. For example, Jupiter is not round. As shown in Figure 7-1, Jupiter is slightly fatter across the equator than through the poles. Indeed, Jupiter's equatorial diameter is about 9,000 kilometers larger than the diameter through the north and south poles.

Jupiter's rapid rotation also plays an important role in determining the planet's overall appearance. Even a small amateur telescope suffices to reveal Jupiter's distinctively striped cloud-tops. As shown in Figures 7-2 and 7-3, the upper atmosphere exhibits a series of light and dark bands that circle the planet parallel to the Jovian equator. By convention, the light-colored whitish stripes are called *zones,* and the darker, reddish-brown stripes are called *belts.*

Anyone who has ever watched a weather report during an evening television newscast is aware that the meteorology of our planet is strongly affected by atmospheric "highs" and "lows." Highs are regions of greater-than-average atmospheric pressure. If we could actually see the air from space, we would notice a bulge in the earth's atmosphere where an exceptional amount of air is piled up. This mound of air weighs down heavily on the earth's surface, thereby producing the high pressure. Similarly, lows are locations of less-

Figure 7-1 Jupiter

Jupiter is by far the largest, most massive, and most rapidly spinning planet in the solar system. Because of its rapid rotation, Jupiter is somewhat flattened at the poles. Even a small amateur telescope suffices to reveal Jupiter's striped appearance and the Great Red Spot. One of Jupiter's moons, Ganymede, is seen along with its shadow on the planet's surface. (Hale Observatories.)

138

Figure 7-2 Belts and Zones
Pioneer 10 flew past Jupiter on December 4, 1973. Prior to closest approach, Pioneer 10 sent back this photograph taken at a distance of 1,840,000 kilometers (1,140,000 miles) from the planet. Numerous details in the dark-colored belts and light-colored zones are easily distinguished. (NASA.)

139

Figure 7-3   The Great Red Spot
Pioneer 11 flew past Jupiter on December 3, 1974. Prior to closest approach, Pioneer 11 sent back this photograph taken at a distance of 1,100,000 kilometers (690,000 miles) from the planet. Three Earths could fit side by side in the Great Red Spot. (NASA.)

than-average atmospheric pressure. They would appear as depressions in the atmosphere where less-than-average amounts of air are pressing down on the earth. Just as water flows downhill, the dominant direction of wind flow is from high-pressure regions (bulges in the atmosphere) toward low-pressure regions (depressions in the atmosphere). Rain and snow, clouds and sunshine come and go as the highs and lows gradually wander across our small, slowly rotating planet.

During the early 1970s, two spacecrafts flew past Jupiter taking numerous photographs and sending back vast quantities of data about the Jovian atmosphere and environment. A typical close-up view of some belts and zones is shown in Figure 7-4. From all this information, it soon became clear that the light-colored zones are atmospheric highs, while the dark-colored belts are atmospheric lows. Warm gases rise upward in a zone and cool off upon reaching the cloud-tops. These cooled gases then spill over into the adjacent low-pressure belts, as shown in Figure 7-5. Due to the planet's rapid rotation, strong trade winds and powerful jet streams flow at the boundaries between the belts and zones. In fact, the major curculation patterns in Jupiter's atmosphere is *exactly* what you would get if you took the highs and lows here on Earth and stretched them out all the way around the planet. In other words, because of Jupiter's rapid rotation, cyclonic and anticyclonic wind-flow associated with highs and lows are wrapped completely around the huge planet. The highs are the zones and the lows are the belts. At extreme latitudes, in the north and south polar regions, this orderly pattern of high-pressure zones and low-pressure belts breaks up into random hurricanes, as shown in Figure 7-6.

In addition to belts and zones, a small telescope also easily reveals a remarkable semipermanent feature in Jupiter's atmosphere called the Great Red Spot. The Great Red Spot was first seen by the Italian astronomer Giovanni Domenico Cassini in 1665. Although the vividness of its color sometimes fades slightly, the feature has been observed for over three centuries.

Like the planet on which it resides, the Great Red Spot is vast. Its width is about 14,000 kilometers (8,700 miles), while its length has varied over the ages between 30,000 to 40,000 kilometers

Figure 7-4 Jupiter's Cloud-Tops
This close-up view of belts and zones was photographed by Pioneer 10 at a range of only 954,000 kilometers (594,000 miles). Intricate details never before seen from Earth are clearly visible. (NASA.)

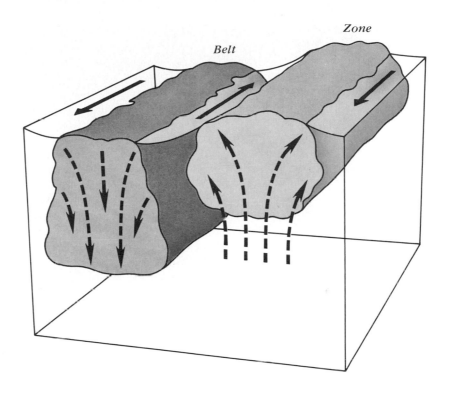

Figure 7-5   Atmospheric Circulation
The light-colored zones are high-pressure regions, while the darker belts are low-pressure regions. Warm gases rise upward through the zones, cool, and spill over into the belts.

Figure 7-6 Turbulence Near the North Pole
*In the polar regions, the familiar pattern of belts and zones
becomes unstable. Circulation breaks up into numerous swirls and
eddies. This view was photographed by Pioneer 11 at a range of
600,000 kilometers (370,000 miles). (NASA.)*

144

(19,000 to 25,000 miles). In other words, three Earths could comfortably fit side by side across the Great Red Spot. An excellent view of the Great Red Spot taken from Pioneer 11 is shown in Plate 7.

The Great Red Spot is located in a prominent zone just south of the Jovian equator called the South Tropical Zone. Data from the flights of Pioneers 10 and 11 have revealed that the rust-colored cloud-tops of the Great Red Spot extend a few kilometers *above* the surrounding cloud-tops of the South Tropical Zone. But as we have seen, the South Tropical Zone is itself a high-pressure region on Jupiter; the cloud-tops of zones are a few kilometers higher than the cloud-tops of belts. Thus, the Great Red Spot is a high-pressure region located in an already-higher-than-average-pressure zone. The Great Red Spot is therefore sometimes called a *superzone.*

Winds blow counterclockwise in the Great Red Spot. Like a vast high-pressure hurricane, the entire Great Red Spot rotates once every 12 earth-days. Recent computer calculations by Andrew Ingersoll have shown that circulation on this scale in Jupiter's atmosphere is comparatively stable and friction-free. In other words, like a wheel rolling between two oppositely moving surfaces (see Figure 7-7), the Great Red Spot rotates without experiencing much viscosity or "drag" from the winds in the surrounding zone. For this reason, the enormous hurricane has been able to rage for over three centuries. If the Great Red Spot were substantially smaller, the amount of "drag" would be correspondingly higher and the hurricane would soon disappear. Indeed, several "small red spots" have occasionally been seen on the planet. These smaller storms (*only* as big as Earth!) peter out after a year or two.

Although scientists are beginning to understand the dynamics of Jupiter's meteorology, the colors and hues seen in the planet's atmosphere remain quite mysterious. As viewed through an amateur telescope or from interplanetary spacecrafts (see Plates 5 and 6), zones are distinctly whitish with, at most, a slight yellowish tinge. In sharp contrast, belts are a rusty orange-brown.

As we have seen, most of Jupiter (that is, 99 percent) consists of hydrogen and helium. Neither of these colorless gases can be responsible for any of the tints and hues that dominate the appearance of the planet's atmosphere. Other substances known to be present in

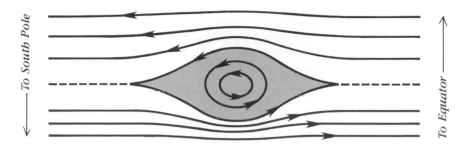

To South Pole

To Equator

Figure 7-7   Winds in the Great Red Spot
Winds blow counterclockwise in the Great Red Spot, causing the storm to rotate once every 12 earth-days. Like a wheel rolling between two oppositely moving surfaces, circulation of the Great Red Spot is comparatively friction-free. Because of the lack of "drag," this high-pressure hurricane has raged for over three centuries. (Adapted from Andrew P. Ingersoll.)

the Jovian atmosphere include methane, ammonia, and water vapor. Although these are also colorless gases, ammonia ice crystals may account for the light-colored cloud-tops of zones. Gases containing ammonia cool rapidly as they rise upward in a zone. Near the cloud-tops, the temperature has declined to a frigid $-130°C$ ($-200°F$), and the ammonia crystallizes into whitish snowflakes.

But what happens to produce the ruddy-colored material as the gases from the zones spill over into the neighboring belts? Some scientists believe that photochemical reactions may occur as certain substances are exposed to the sun's ultraviolet rays at Jupiter's cloud-tops. The ultraviolet light simply causes a reaction that converts these yet-to-be-discovered chemicals into a reddish-brown substance. As the murky material descends into the belts, it is dissociated back into its components only to reemerge once again—clean and colorless—rising back up through the zones.

But there is a second explanation — an explanation that involves the fascinating possibility of biology.

In the early 1950s, Harold Urey and Stanley Miller performed a classic experiment whereby a gaseous mixture of ammonia, methane, water vapor, and hydrogen was placed in a container and subjected to electric sparks, The two chemists were simply trying to simulate lightning discharges in the Earth's primordial atmosphere. After several days, the flask was found to contain amino acids. Amino acids are the molecular building blocks of proteins.

This experiment has been performed and studied many times, most recently by Cyril Ponnamperuma at the Laboratory of Chemical Evolution at the University of Maryland. Although we are far from being able to create life directly from inorganic chemicals, it is clear that many of the molecular building blocks of living matter are produced in these experiments. Remarkably, the murky brew that is manufactured by these experiments has a distinctive reddish-brown color — exactly the same hues and tints seen in the belts of Jupiter.

It seems entirely reasonable to suppose that these same chemical reactions are occurring in Jupiter's turbulent atmosphere. In 1955, radio astronomers discovered that powerful, intermittent bursts of static were coming from Jupiter. The obvious explanation for this crackling static is that lightning bolts are constantly flashing among the Jovian clouds and thunderheads. The chemicals are there; the electric sparks are there. So why not amino acids and nucleotides? And since this chemical processing has been occurring for billions of years (rather than just a few days, as in laboratory experiments), perhaps enough time has elapsed for the development of living organisms. At a depth of only 100 kilometers below the Jovian cloud-tops, temperatures and pressures are quite hospitable — even by terrestrial standards. Simple one-celled creatures could easily survive, floating among the clouds at these depths.

In view of the negative results from the Viking biology experiments on Mars, the Jovian clouds seem to be the next obvious target in the search for extraterrestrial life. Ambitious missions are now being planned for the 1980s. Probes will be dropped by parachute into the Jovian atmosphere from orbiting satellites. During descent, these miniature chemical/biological laboratories will transmit data to the satellites overhead for relay back to Earth. Analysis of these data may provide important clues concerning the origin of life here on our own planet. After all, the chemical composition of Jupiter's atmosphere today is generallly believed to be the same as Earth's primordial atmosphere 4 billion years ago.

In 1977, Carl Woese of the University of Illinois announced his identification of an extraordinary, previously unidentified form of life here on Earth. Unlike bacteria or plant and animal life, this newly discovered life form exists only in oxygen-free environments such as the deep hot springs of Yellowstone National Park and at the ocean floor. These simple organisms superficially resemble bacteria, but they take in carbon dioxide, water, and hydrogen and give off methane. It is quite possible that these one-celled organisms are directly related to the primordial organisms from which *all* life on Earth eventually evolved. These ancient organisms are ideally suited to Earth's primordial atmosphere. And, in addition, they would thrive in Jupiter's clouds. Indeed, it is tantalizing to speculate that some of the methane (often called "swamp gas" or "marsh gas") we detect in the Jovian atmosphere might actually be caused by biological processes in the belts and zones.

In designing spacecrafts for missions to Jupiter, engineers must take great care to shield their instruments against the intense radioactivity arising from particles trapped by Jupiter's magnetic field. Like Earth, Jupiter has a magnetic field that captures charged particles (protons and electrons) from the solar wind. In the case of our planet, these particles are trapped in two huge, doughnut-shaped belts (called the Van Allen radiation belts) that completely surround the Earth, as shown in Figure 4-6. These Van Allen belts dominate the inner portions of Earth's magnetosphere. The outer portions are strongly affected by interactions with the solar wind. A shock wave is formed where the supersonic solar wind is abruptly slowed by its first encounter with the geomagnetic field. Inside this so-called *bow shock* there is a turbulent region, the *magnetosheath,* where the solar wind tries to flow around the Earth. The inner boundary of the magnetosheath is called the *magnetopause,* where the pressure from the solar wind is counterbalanced by the magnetic pressure of the

Earth's field. Due to the constant blowing of the solar wind, Earth's magnetosphere is swept outward, away form the sun, like a giant invisible comet.

Astronomers have known for a long time that Jupiter must possess a strong magnetic field. A planetwide magnetic field provides the simplest explanation for continuous radio noise first detected from Jupiter in the late 1950s. Much of this uninterrupted static can be easily explained by high-speed electrons moving through the planet's magnetic field. But a deeper understanding of Jupiter's magnetosphere had to wait for the flights of Pioneers 10 and 11 in the early 1970s. These historic missions revealed that Jupiter's magnetic field is roughly ten times as strong as Earth's. In addition, Jupiter's magnetic field is upside down compared with Earth's. On Jupiter, a scout's compass would point toward the south pole.

The Earth's magnetic field is thought to arise from electric currents inside the liquid portion of our planet's iron core. Like a giant dynamo, the Earth's rotation causes these currents to produce our planet's overall magnetic field. But, of course, Jupiter does not possess an iron core. After all, the planet is composed mostly of hydrogen. So what could be the source of the enormous Jovian magnetic field?

Far below Jupiter's cloud-tops, the atmospheric pressure must be enormous. The crushing weight of trillions upon trillions of tons of gas pressing downward give rise to conditions undreamed of here on Earth. Indeed, nearly halfway to Jupiter's core, pressures are so great that liquefied hydrogen is actually turned into metal. Although liquid metallic hydrogen has never been observed in laboratories, scientists feel so sure of their understanding of hydrogen atoms that they confidently predict the existence of liquid metallic hydrogen at extreme pressures. As shown in Figure 7-8, astronomers believe that most of Jupiter's interior consists of liquid metallic hydrogen. Electric currents in this liquid metallic interior are probably the source of Jupiter's magnetic field, just as currents in Earth's liquid iron core produce the geomagnetic field.

Since Jupiter's magnetic field is so much stronger than Earth's, the Jovian magnetosphere is truly enormous. If you could see Jupi-

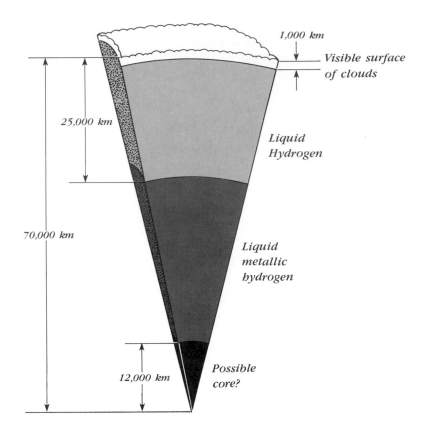

Figure 7-8   The Interior Structure of Jupiter
Jupiter consists mostly of hydrogen. Liquid hydrogen dominates the planet's structure down to a depth of 25,000 kilometers beneath the cloud-tops. But below this level, where the pressure exceeds 3 million earth-atmospheres, the liquid hydrogen is converted into its metallic form.

ter's magnetosphere at night with your naked eyes, it would appear sixteen times as large as the full moon, even though the planet itself looks like nothing more than a very bright star.

As in the case of Earth, Jupiter's magnetosphere is surrounded by a bow shock, a magnetosheath, and a magnetopause, as shown in Figure 7-9. But because of the planet's rapid rotation, charged particles inside the Jovian magnetosphere experience considerable centrifugal forces. Thus, instead of being trapped in doughnut-shaped radiation belts, the particles are flung out into a huge sheet, called the *current sheet,* that lies roughly parallel to the planet's magnetic equator.

An important complication affects Jupiter's vast magnetosphere that is simply not relevant to our own planet. Five of Jupiter's fourteen moons have orbits *inside* the Jovian magnetosphere. All four of the giant Galilean satellites (Io, Europa, Ganymede, and Callisto) as well as tiny Amalthea have orbits so close to Jupiter that they are constantly sweeping up charged particles, leaving behind empty corridors in the magnetosphere. This causes temporary reductions in the radiation near Jupiter; there is even some evidence that the entire magnetosphere is drained of particles every ten hours.

Studying Jupiter's magnetosphere is certainly one of the major goals of the 1980s. Of course, this would be best done by planet-orbiting spacecrafts rather than by the brief flybys of Pioncers 10 and 11 and Voyagers 1 and 2. A significant bonanza of these future missions would be the opportunity to study the Galilean satellites from close range. As we saw in Chapter 5 (see Figure 5-1), each of these four moons is as big as Mercury. They are all really planets in their own right. And each seems to have its own distinctive character. Io is composed mostly of rock. Europe is probably covered with a thick layer of ices. And Ganymede and Callisto may also possess substantial subsurface deposits of water. Future missions to these worlds will probably reveal huge snowdrifts, frozen oceans, vast canyons and cratered landscapes, volcanoes, and geysers of liquid ammonia and methane. Based on previous experience, we shall also probably find many features that were never thought of, even in our wildest imaginings.

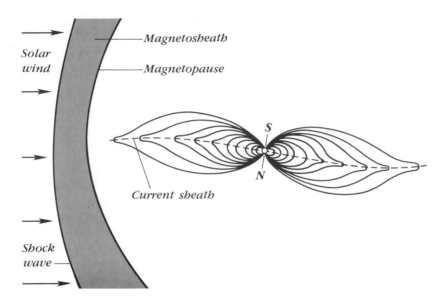

Solar
wind

—Magnetosheath

—Magnetopause

S

N

Current sheath

Shock
wave

Figure 7-9   Jupiter's Magnetosphere
*Jupiter's magnetic field is roughly ten times as strong as Earth's. The
dimensions of Jupiter's magnetosphere are therefore approximately
a hundred times as large as Earth's. Because of Jupiter's rapid
rotation, charged particles are confined to a huge sheet parallel to
the planet's magnetic equator.*

Figure 7-10   A Volcano on Io
*Voyager 1 glimpsed this volcano near Io's limb during the flyby in
March 1979. It is possible that volcanoes on Io contribute debris
to the ring around Jupiter. (NASA.)*

Figure 7-11   The Surface of Ganymede
Voyager 1 took this close-up photograph of Ganymede at a range of only 145,000 kilometers (87,000 miles) on March 5, 1979. The distance across the bottom of the picture is 580 kilometers (350 miles). Notice the strange system of grooves and channels that criss-cross Ganymede's icy crust. (NASA.)

153

8

The Outer Worlds

Saturn is surely among the most spectacular sights that can be seen through a telescope. Although dramatic color photographs of nebulas and galaxies are found in most astronomy books, direct telescopic views of these distant objects in the universe are often disappointing. Even through the most powerful telescopes, a nebula or a galaxy simply looks like a faint, hazy, grayish blurr. All those breathtaking color photographs are the result of long time exposure photography that picks up traces of hues and tints far below the sensitivity of the human eye.

But none of these problems plague views of Saturn. A glance through the telescope's eyepiece reveals a bright, unearthly world surrounded by a dramatic system of rings. Along with craters on the moon and markings on Mars, Saturn is one of the standard favorites of backyard astronomers with inexpensive telescopes. The sight is quite similar to Figure 8-1. Novices getting their first view of Saturn might complain that the planet appears smaller than they had anticipated. They need only be reminded that Saturn is nearly a billion miles away.

In many respects, Saturn takes second place to Jupiter. Saturn is the second largest, second most massive, and second most rapidly rotating planet in the solar system. For comparison, data for all four Jovian planets are given in the table on page 158. Saturn's gaseous surface even exhibits a striped appearance similar to its giant neighbor. But in keeping with its second-class status, Saturn's belts and zones are not as pronounced as those on Jupiter. No features comparable to the Great Red Spot have ever been seen on Saturn.

Of course, the most outstanding attribute of the sixth planet form the sun is a magnificent system of rings. Saturn's rings were first seen in 1655 by the Dutch astronomer Christian Huygens (the same fellow who produced the first crude map of Mars). Within a few years, astronomers realized that this remarkable feature about Saturn actually consists of several concentric rings. In fact, four rings are generally recognized today.

The first hint of several rings came in 1675, when the Italian astronomer Giovanni Cassini (the same fellow who produced the first drawings of the Great Red Spot on Jupiter) discovered a gap in

Figure 8-1   Saturn

Saturn is the sixth planet from the sun. At a distance of 1½ billion kilometers (nearly 900 million miles) from the sun, Saturn takes 29½ years to complete one orbit. The planet's beautiful rings can be easily seen through amateur telescopes. (New Mexico State University Observatory.)

	Diameter (Earth = 1)	Mass (Earth = 1)	Period of rotation
Jupiter	11.2	317.9	9^h50^m to 9^h55^m
Saturn	9.5	95.2	10^h14^m to 10^h38^m
Uranus	3.7	14.6	Approximately 11^h
Neptune	3.5	17.2	Approximately 16^h

the rings. This gap, called *Cassini's division,* is easily seen through amateur telescopes or on any good-quality photograph. Cassini's division separates the Outer Ring (also called Ring A) from the Bright Ring (also called Ring B). The gap is about 5,000 kilometers (roughly 3,000 miles) wide.

The third ring, called the Crepe Ring or Ring C, was first noticed in 1850. The Crepe Ring lies immediately inside Ring B and shows up only faintly on the finest photographs. In 1969, some observers reported discovering a fainter, innermost ring. This Ring D lies immediately inside Ring C and extends down to within 12,000 kilometers (7,500 miles) of the planet's gaseous surface.

The ring system is huge. The outer diameter of the outermost ring is 274,000 kilometers (170,000 miles). That is nearly *twice* the diameter of Jupiter. But in spite of their vast extent, the rings are exceptionally thin. Background stars can easily be seen shining through the rings. In addition, because of their orientation as viewed from Earth, the rings are occasionally seen edge-on. During these occasions, the rings seem to vanish. Astronomers therefore surmise that the rings are no more than 5 kilometers (3 miles) thick.

By analyzing reflected sunlight from the rings, astronomers easily realized that the rings are not a solid, continuous sheet like a sheet of metal. Instead, the rings consist of countless trillions of pebble-sized rocks and chunks of ice. Just as Mercury revolves about the sun much more rapidly than Pluto, pebbles in the innermost ring revolve about Saturn more swiftly (once every 2 hours) than pebbles at the outer edge of the Outer Ring (once every 15 hours).

Figure 8-2   Saturn
Like Jupiter, Saturn is slightly flattened because of its rapid rate of rotation. The equatorial diameter (120,000 kilometers, or 75,000 miles) is slightly larger than the diameter through the poles (107,000 kilometers, or 66,000 miles). (Hale Observatories.)

159

All of these pebbles are probably coated with a layer of frost. Evidence for dry ice frost in Saturn's rings dates back to infrared observations by Gerard P. Kuiper in 1949 and 1957. A frosty coating would help explain why the rings are so highly reflective and can be seen so easily through telescopes even though only a small amount of matter is involved.

The total mass of Saturn's rings is only about a hundredth of the mass of our moon. In other words, if all the material in Saturn's rings were lumped together, the resulting object would be the size of one of the typical *small* moons in the solar system. In fact, the rings consist of rock and ice that did not manage to accrete and coalesce into a satellite. It is generally believed that all major satellites formed from the gradual accumulation of particles that orbited the proto-planets when the solar system was in its infancy. But in Saturn's case, the particles in the rings are simply too close to the planet. The tidal forces due to Saturn's gravity across the rings are stronger than the gravitational forces between the individual particles in the rings. Therefore, the stronger tidal forces — forces that characteristically try to pull things apart or stretch things out — simply prevented the natural formation of yet another satellite.

Saturn's rings lie in the plane of Saturn's equator. But like Earth, the plane of Saturn's equator is tilted out of the plane of the planet's orbit about the sun. In the case of Earth, the angle between the equator and the orbit is 23½°. This tilt is responsible for the seasons on our planet. In the case of Saturn, this angle happens to be 27°. Of course, this means that Saturn also experiences seasons. But it also means that the appearance of Saturn as seen from Earth changes dramatically over the years. As diagramed in Figure 8-3, earth-based observers occasionally look "down" on the rings, and, half a Saturnian year later, we see the rings from "underneath." A series of photographs showing the changing appearance of Saturn is displayed in Figure 8-4. Notice that the rings seem to disappear when viewed edge-on.

Saturn's rings were last viewed edge-on in 1966. During that time, the substantial reduction of glare from the rings permitted the discovery of Saturn's tenth moon, Janus. Janus is Saturn's innermost moon and orbits the planet just beyond the outer edge of the Outer

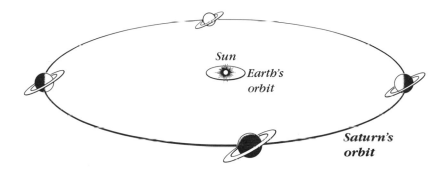

Figure 8-3   The Tilt of Saturn's Rings
*Saturn's rings are tilted by 27° from the plane of the planet's orbit
about the sun. Earth-based observers therefore view the rings at
various orientations as Saturn orbits the sun.*

Figure 8-4   Saturn's Changing Appearance
*Earth-based observers see the rings at various inclinations as
Saturn orbits the sun. Every 15 years we have the opportunity to
view the rings edge-on. (Lowell Observatory.)*

Ring. Janus is a very tiny satellite, and its discovery cannot be reconfirmed until the rings are again seen edge-on in 1981. Of course, Janus might be glimpsed by one of the three spacecrafts (Pioneer 11, Voyager 1, and Voyager 2) that will fly past Saturn between September 1979 and August 1981.

Of the ten satellites in orbit about Saturn, the largest moon, Titan, was first seen by Huygens in 1655. Titan is one of the seven largest moons in the solar system. It is bigger than any of the fourteen moons that circle Jupiter and may even be a few miles larger in diameter than the planet Mercury. Titan is so large that its gravity has been strong enough to retain an atmosphere. In 1944, Kuiper discovered evidence of methane gas during a spectroscopic analysis of reflected sunlight from the satellite. Other gases are also believed to be present.

During the late 1600s, Cassini discovered four more satellites: Tethys, Dione, Rhea, and Iapetus. Of these, Iapetus is especially intriguing. The brightness of Iapetus varies dramatically as it orbits Saturn. Evidently one side of Iapetus is very highly reflective, while the other side is dark and dull. Iapetus was the final target in the famous science fiction novel *2001: A Space Odyssey* by Arthur C. Clarke. Unfortunately, this fact was not carried through to the screenplay version of the story. Nevertheless, we may soon view this world during three flybys that will occur two decades *before* the twenty-first century.

Of the remaining satellites, the outermost is Phoebe. Phoebe is at an average distance of 13 million kilometers (8 million miles) from Saturn and takes 550 days to complete one orbit. The overwhelming majority of satellites in the solar system orbit their planets in the same direction that the planets rotate, which is the same direction that the planets revolve about the sun. But Phoebe is one of the few retrograde satellites. It is going around Saturn backward.

Of the inner Saturnian satellites, Mimas has a direct effect on the structure of the rings. Mimas orbits Saturn every 22.6 hours. But recall that particles in Saturn's rings orbit the planet with periods ranging from 2 to 15 hours. In particular, if there were any pebbles inside Cassini's division, they would have an orbital period of 11.3 hours. That is exactly half of Mimas's period.

This explains the existence of Cassini's division. If there were any particles inside Cassini's division, they would line up with Mimas every two orbits about the planet. Thus, every 22.6 hours, Mimas would exert the same gravitational pull in the same direction on each of these particles. Although Mimas is tiny and each individual gravitational tug is very weak, repeated action over millions of years will cause the particles to deviate from their original 11.3-hour orbits. Any particles that start off with a 11.3-hour orbit eventually end up with a different orbit somewhere else in the rings. The final result is a gap that we today call Cassini's division.

Until recently, it was believed that ringed Saturn was unique among the planets. But now we know of a second planet that possesses a system of rings. The rings were discovered accidently just as the planet itself was discovered by chance.

On March 13, 1781, William Herschel was viewing a portion of the constellation of Gemini through his telescope when he noticed an unusual star that was "visibly larger than the rest." Herschel initially thought that he had discovered a distant comet. But subsequent observations soon revealed that the object has a nearly circular orbit far outside the orbit of Saturn. Herschel had stumbled upon Uranus, the seventh planet from the sun.

Although Uranus was discovered with a telescope, the planet does occasionally become bright enough to be viewed with the naked eye. But you must have superb eyesight, you must be located under a clear moonless sky, and you must know exactly where to look. At best, Uranus is as bright as the dimmest stars that the naked eye can perceive. In fact, Uranus was plotted as a star on at least 20 star charts prior to Herschel's discovery.

Through a telescope, Uranus is a ghostly greenish-blue world surrounded by five moons. All of these satellites are rather small; three are shown in Figure 8-5. Although some observers have reported seeing faint markings on the planet, no surface details have ever been photographed. An extraordinarily fine photograph of Uranus was taken from a balloon-borne telescope at 80,000 feet above the earth in 1970. The result, a composite of more than a dozen separate images, is shown in Figure 8-6. Although the resolu-

Figure 8-5 Uranus
Uranus is the seventh planet from the sun. At a distance of nearly 3 billion kilometers (1¾ billion miles) from the sun, Uranus takes 84 years to complete one orbit. Uranus has five small moons, three of which are identified in this photograph. (Lick Observatory.)

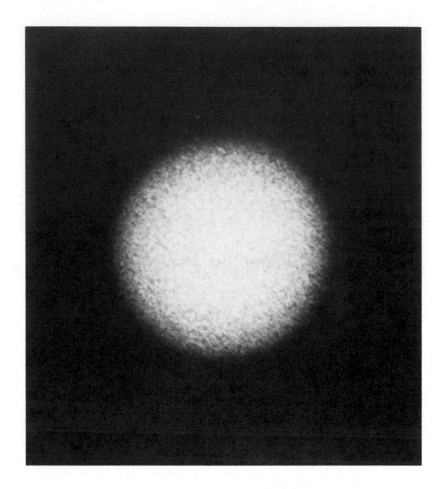

Figure 8-6   The Best View of Uranus

This exceptionally clear image of Uranus was obtained from a balloon-borne telescope 80,000 feet above the Earth's surface. Notice that the planet is totally featureless. Continent-sized markings as small as 2,000 kilometers (1,200 miles) across should have been detectable. (Project Stratoscope II, Princeton University, supported by NSF and NASA.)

tion was good enough to reveal continent-sized features, no markings at all were recorded. Better views of Uranus should come from the flyby of Voyager 2 in January 1986.

One unusual property of Uranus is that it is oriented very differently from any of the other planets. All the other planets have their axes of rotation approximately perpendicular to the planes of their orbits. Slight tilts ($23\frac{1}{2}°$ for Earth, $24°$ for Mars, $3°$ for Jupiter, and $27°$ for Saturn) are responsible for seasons. But Uranus's axis of rotation is tilted through $98°$; Uranus's axis is practically *in* the plane of its orbit, as shown in Figure 8-7. This means that the seasons on Uranus are strange. For example, during summer in the northern hemisphere, the sun appears suspended for years nearly overhead at the North Pole. And when winter comes, nearly half the planet is plunged into a frigid winter "night" that lasts for decades. Bizarre weather patterns surely result from this unusual and uneven heating of the planet's atmosphere.

Uranus was scheduled to pass in front of a faint star on March 10, 1977. This event, called an *occultation,* is important to astronomers because it affords the opportunity to measure the planet's diameter with substantial accuracy. From knowing how fast the planet is moving along its orbit, and from measuring how long the star's light is blocked out, the planet's diameter can be easily deduced.

Calculations had shown that the occultation would be best seen over the Indian Ocean. Flying with all their equipment onboard a NASA airplane, a team of astronomers from Cornell University patiently waited. It was all supposed to be very routine. But half an hour before the occultation was scheduled to begin, the star briefly disappeared from view *five times.* The occultation came and went. And then, half an hour after the occultation, the star blinked on and off five more times. The meaning of this totally unexpected result was instantly clear to the ecstatic airborne astronomers: Uranus is surrounded by five thin rings.

Uranus's rings are very different from those that surround Saturn. Saturn's rings are very bright and very broad. In sharp contrast, Uranus's rings are dark and thin. Unlike the highly reflective frost-covered pebbles in Saturn's rings, the rocks in Uranus's rings

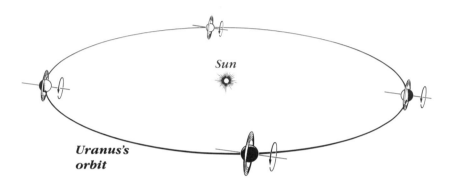

Sun

*Uranus's
orbit*

Figure 8-7   Uranian Seasons
*Uranus's axis of rotation is tilted so far that it nearly lies in the
plane of the planet's orbit. This unusual orientation produces
strange seasons and bizarre weather. For example, during summer
in the northern hemisphere, the sun is suspended nearly overhead at
the north pole.*

must have the reflectivity of lumps of coal. If this were not so, the rings would have been noticed in earlier observations. In addition, Uranus's five rings are very thin and widely spaced. The inner rings are perhaps only 10 kilometers wide, and the thicker, outer ring has a width of roughly 100 kilometers. The innermost ring is located 18,000 kilometers (11,000 miles) above Uranus's cloud-tops, while the outer ring is at an altitude of 25,000 kilometers (16,000 miles) above the planet.

Historically, Uranus played an important role in nineteenth-century science. During the early 1800s, it was noticed that Uranus was not exactly following its predicted path. Up to this time, classical Newtonian mechanics had been used to calculate the orbits and positions of the planets and their satellites with unprecedented accuracy. But here was a new planet that did not seem to follow the rules. Could it be that the classical laws of physics simply do not work that far from the sun?

While the faith of some scientists was being shaken, two astronomers independently realized that Uranus's abnormal behavior could be explained by effects of a more distant, yet-undiscovered planet. John Couch Adams in England and Urbain Jean Joseph Leverrier in France both reasoned that the small deviation of Uranus from its predicted orbit might be caused by the gravitational pull of an unknown distant planet. Their calculations gave a suspected location for the unknown planet. And on September 23, 1846, J. G. Galle at the Berlin Observatory received a letter from Leverrier giving the location. That very night Galle sighted Neptune, the eighth planet from the sun, less than 1° away from the predicted position.

Neptune was literally discovered with pencil and paper. Classical mechanics and Newton's laws are so powerful and so universal that they can be used to predict the existence of undiscovered planets!

In many respects, Uranus and Neptune are twins. They are nearly the same size, have nearly the same mass and density, and probably possess nearly the same chemical composition and internal structure. But Neptune is so incredibly remote that few details about the planet are known. Viewing Neptune from Earth is like examining a dime from a distance of one mile.

Figure 8-8 Neptune
Neptune is the eighth planet from the sun. At a distance of 4½ billion kilometers (2¾ billion miles) from the sun, Neptune takes 165 years to complete one orbit. The arrow in this photograph points to Triton, the largest of Neptune's two moons. (Lick Observatory.)

Neptune is orbited by two moons, one of which is exceptionally large. In fact, Triton is the largest satellite in the entire solar system. Triton's diameter is 6,000 kilometers (almost 4,000 miles). For comparison, the diameter of Mercury is 4,880 kilometers (3,030 miles), and our own moon measures only 3,476 kilometers (2,160 miles) across.

Triton is unusual. Triton is one of those rare retrograde satellites. It goes around Neptune backward. And Neptune's other satellite, Nereid, also has an unusual orbit. Nereid's orbit is more highly elliptical than that of any other satellite in the solar system. These peculiarities suggest that something extraordinary might have happened in Neptune's past. This notion is reinforced by considering Pluto.

Pluto was discovered in February 1930 by Clyde Tombaugh at Lowell Observatory as the result of a lengthy photographic search. It was soon apparent that Pluto does not fit into the scheme of the outer Jovian planets. Pluto is very tiny. It is even smaller than Mercury. In addition, Pluto's orbit is more highly elliptical than that of any other planet. Indeed, the orbit is so eccentric that Pluto is occasionally closer to the sun than Neptune is.

In June 1978, James W. Christy of the U.S. Naval Observatory was examining several high-quality photographs of Pluto when he noticed something unusual. On some of the photographs, Pluto seemed to be elongated, as though it had a small lump on one side. After scrutinizing earlier photographs that also occasionally showed a lump, Dr. Christy and his colleagues came to the only logical conclusion: they had discovered a moon going around Pluto. The satellite was named Charon, after the mythical ferryman who transported souls across the river Styx to the underworld ruled by Pluto. A photograph of Pluto and Charon is given in Figure 8-10.

Charon orbits Pluto once every 6 days, 9 hours, and 17 minutes. This is exactly the same time it takes for Pluto to turn once about its axis. Consequently, Pluto perpetually keeps the same side facing Charon. This unique situation of synchronous rotation and revolution suggests that Pluto is not round. Perhaps the planet is somewhat egg-shaped.

Figure 8-9   Pluto
Tiny Pluto is so far away that it looks just like a faint star. Pluto can be identified by comparing photographs taken a few days apart and looking for a moving object. (Lick Observatory.)

Because of the peculiarities of the orbits of Triton and Nereid about Neptune as well as the unusual orbit of Pluto around the sun, some people have suggested that Pluto might be an escaped moon of Neptune. Indeed, Pluto's low density means that it is composed of iccs, likc most of the other satellites of the outer planets. Perhaps some catastrophic event happened in Neptune's distant past that severely disrupted the orbits of Neptune's moons, causing one of them to be ejected completely.

Early in 1978, R. S. Harrington and T. C. Van Flandern of the U.S. Naval Observatory announced the results of an intriguing calcu-

lation. They start off assuming that Triton, Nereid, and Pluto once circled Neptune in rather ordinary orbits. They then assume that a very large object passed near Neptune. Of course, this object would disrupt the paths of Neptune's moons causing them to go into new orbits. Harrington and Van Flandern reconstructed the scenario with a computer and discovered that "orbits very much like those of Pluto, Triton, and Nereid can result from such a close encounter of a massive body with Neptune."

Following the discovery of Charon later in 1978, Harrington and Van Flandern argued that a satellite in synchronous orbit around

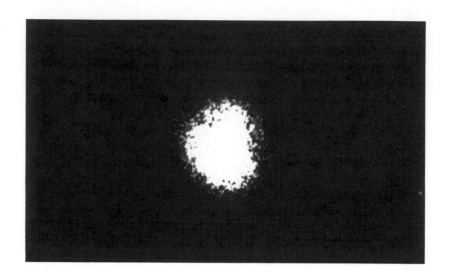

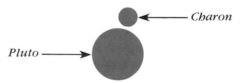

Figure 8-10   Pluto and Charon

The distance between Pluto and its moon is only 20,000 kilometers (12,000 miles). Consequently, the photographic images of these two tiny, remote worlds are blurred together. This enlargement was made by Stuart Jones of Lowell Observatory. (U.S. Naval Observatory.)

Pluto favors their theory. The same violent event that yanked Pluto out of its ancient orbit around Neptune also pulled Pluto apart, causing it to break into two pieces.

Did such an encounter ever occur? Important clues might come from Voyager 2 as it flys past Neptune in September 1989 — if the spacecraft lasts that long. Otherwise, the mystery will remain hidden in the remote depths of interplanetary space.

9

Interplanetary Vagabonds

At first glance, the arrangement of planetary orbits in the solar system seems quite haphazard. Some planets are crowded close to the sun, whereas others are widely scattered in immense orbits. Astronomers usually prefer to measure the sizes of these orbits in *astronomical units.* By agreement, one astronomical unit (abbreviated AU) is the average distance between the earth and the sun (about 150 million kilometers, or 93 million miles). Thus, for example, the average distance between Mercury and the sun is 0.39 AU, whereas Pluto is a hundred times as far away, at 39.5 AU from the sun.

During the 1770s, Johann Bode, director of the Berlin Observatory, popularized a simple mathematical scheme for remembering the distances of the planets from the sun. First, write down the sequence of numbers 0, 3, 6, 12, 24, ..., doubling each time. Then add 4 to each number. Then divide each of the sums by 10. As shown in the table below, the final sequence of numbers is remarkably close to the sizes of planetary orbits expressed in astronomical units.

This scheme is often called Bode's law, in spite of the fact that it was neither invented by Bode nor is it a "law." Nevertheless, Bode's law received great notoriety when William Herschel discovered Uranus in 1781. Notice that the actual distance between Uranus and the sun is very near the "predicted" distance according to Bode's law.

The discovery of Uranus led many people to believe that there was something quite special about Bode's law, beyond a simple scheme to remember planetary distances. In particular, some astronomers began wondering if anything special might exist at 2.8 AU from the sun. Indeed, a team of German astronomers began searching the skies for an object that might be located in the gap between the orbits of Mars and Jupiter.

On January 1, 1801, the Sicilian astronomer Giuseppe Piazzi noticed a faint "star" in the constellation of Taurus that was not drawn on his star chart. Piazzi observed that the "star" gradually moved from night to night, clearly indicating that he had discovered an object in our solar system. By the end of 1801, it was established that Piazzi's object orbits the sun every 4.6 years at an average distance of 2.77 AU, remarkably close to Bode's "predicted" distance. Piazzi had discovered the first *asteroid,* which was named Ceres, after the protecting goddess of Sicily.

Bode's law	Planet	Actual distance (in AU)
(0 + 4)/10 = 0.4	Mercury	0.39
(3 + 4)/10 = 0.7	Venus	0.72
(6 + 4)/10 = 1.0	Earth	1.00
(12 + 4)/10 = 1.6	Mars	1.52
(24 + 4)/10 = 2.8	?	
(48 + 4)/10 = 5.2	Jupiter	5.20
(96 + 4)/10 = 10.0	Saturn	9.54
(192 + 4)/10 = 19.6	Uranus	19.18
(384 + 4)/10 = 38.8	Neptune	30.06
(768 + 4)/10 = 77.2	Pluto	39.53

In March 1802, Heinrich Oblers, a member of the German search team, discovered a second faint, starlike object that turned out to be the second asteroid. It was named Pallas and also orbits the sun every 4.6 years at an average distance of 2.77 AU.

At maximum brightness under the best conditions, both Ceres and Pallas are slightly fainter than the dimmest stars that the naked eye can perceive. You always need a telescope to see asteroids. From this fact, nineteenth-century astronomers immediately realized that Ceres and Pallas were both extremely tiny. Indeed, Ceres is the largest asteroid and has a diameter of only 1,000 kilometers (600 miles). Thus, neither Ceres nor Pallas could qualify as Bode's "missing planet." But since both asteroids have nearly identical orbits, it soon became popular to suppose that Bode's planet had blown up. The asteroids were the remaining fragments.

This exploded-planet hypothesis received support with the discoveries of the asteroids Juno (in 1804) and Vesta (in 1807). Juno orbits the sun every 4.4 years at an average distance of 2.67 AU.

Vesta's 3.6-year orbit is slightly smaller; the distance to the sun is only 2.36 AU. Presumably, these were supposed to be more pieces of Bode's planet.

Discovering asteroids was a difficult and painstaking business. With saintly patience, astronomers scrutinized the heavens for faint, uncharted stars that shift their positions from night to night. For example, the Purssian astronomer Karl Hencke searched for *fifteen years* before he discovered the fifth asteroid, Astraea, in 1845. By 1890, a total of only 300 asteroids had been sighted and catalogued.

The advent of photography had a dramatic effect on the asteroid business. No longer was it necessary to peer for hours through an eyepiece. Instead, simply take a time exposure photograph. If any asteroids happen to be in the field of view, their images will leave blurred trails that are easily distinguished from the images of stars. An excellent example is shown in Figure 9-1.

About 2,000 asteroids have been formally discovered and had their orbits calculated. And surely thousands more have left trails on countless photographs at observatories around the world. But professional astronomers would have to abandon (or at least modify) their research projects to follow asteroids that accidentally appeared on their photographic plates. An asteroid is not officially "discovered" until a reliable orbit has been calculated. And that takes several careful observations on a number of nights. These days, it just doesn't seem to be worth the effort.

All of the big asteroids were discovered long ago. About 230 of them have diameters greater than 100 kilometers (60 miles). But there must be tens of thousands of asteroids with diameters of a few kilometers or less. Some recent estimates of these small asteroids have put their numbers near 100,000.

The vast majority of asteroids circle the sun between the orbits of Mars and Jupiter. This region of the solar system is therefore called the *asteroid belt*. Although asteroids are very numerous, their total mass is very small. All the asteroids lumped together would barely produce a typical small-sized moon — certainly nothing approaching planetary dimensions.

The old exploded-planet hypothesis keeps cropping up even today in science fiction stories. There are severe problems with this

Figure 9-1   Two Asteroids
The blurred trails of two asteroids appear on this time exposure photograph of a star field. Many asteroids are discovered accidently by astronomers photographing various objects in the sky. (Yerkes Observatory.)

Figure 9-2   Phobos
*Most astronomers believe that a typical asteroid would look
like Phobos, the inner satellite of Mars. Phobos is about 20
kilometers (12 miles) in diameter. This photograph was taken from
a distance of 480 kilometers (300 miles) by Viking Orbiter 1.
Features as small as 20 meters (65 feet) can be seen. (NASA.)*

theory. Indeed, there simply isn't enough material in the asteroid belt to make a decent-sized planet in the first place. It is perhaps far more reasonable to suppose that the asteroids are objects that condensed out of the primordial solar nebula but never managed to accrete and coalesce into a planet. Like the trillions of pebbles in Saturn's rings that never succeeded in forming a satellite, countless thousands of boulders in the asteroid belt never made it past the initial stages of planet formation.

This analogy with Saturn's rings is more than superficial. Recall that Cassini's division is caused by the gravitational influence of one of Saturn's inner moons, Mimas. In exactly the same way, Jupiter has produced some gaps in the asteroid belt. Jupiter orbits the sun every 11.86 years at an average distance of 5.20 AU. If an asteroid orbited the sun with a period that is a simple fraction (½, ⅓, etc.) of Jupiter's period, then Jupiter and the asteroid would regularly and frequently line up. For example, an asteroid 3.28 AU from the sun would have a period of 5.93 years (exactly half of Jupiter's) and would pass between the sun and Jupiter in exactly the same location on every second orbit. Continuous action of Jupiter's strong gravity during each of these alignments would eventually cause the tiny asteroid to deviate from its initial orbit. A gap in the distribution of asteroids is left behind.

There are seven major gaps in the asteroid belt at distances from 2.2 to 3.3 AU. They are called *Kirkwood gaps,* after the American astronomer Daniel Kirkwood, who first noticed their existence in 1866. Each gap corresponds to a simple fraction of Jupiter's period.

Asteroids are made of rocky material that sometimes has a substantial iron content. Analysis of reflected sunlight from asteroids reveals that they probably have a chemical composition identical to certain types of meteorites. Meteorites are small interplanetary rocks (called *meteoroids* when they are in space) that have survived a fiery descent through the earth's atmosphere. Most of the rocks that collide with our planet burn up in the atmosphere, as shown in Figure 9-3. But occasionally, surviving fragments (called *meteorites* when you can touch them) are found on the ground.

All meteorites can be classified into three broad categories based on their iron content. First of all, there are stony meteorites

Figure 9-3   A Meteor
*A meteor is a streak of light that occurs when a tiny rock (called a
meteoroid) strikes the earth's atmosphere at high speed. Most
meteoroids are not bigger than a grain of sand and are completely
vaporized in less than a second. (Yerkes Observatory.)*

(also called *stones*) that have the lowest iron content, typically less than 20 percent. As their name suggests, stony meteorites are made of essentially the same material as ordinary rocks on the ground. These meteorites are therefore very difficult to find. Astronomers believe that the vast majority of meteoroids that strike the earth — more than 90 percent — are stones. But if a stony meteorite is allowed to weather under the influence of rain, wind, and snow, it is soon indistinguishable from other rocks on the ground. Stony meteorites, such as the specimen in Figure 9-4a, are thus very valuable and highly prized among collectors.

A second major class of meteorites is called *stony-irons.* As their name suggests, these meteorites are roughly a 50/50 mixture of minerals and iron metal. Sometimes this mixture is quite coarse, as shown in Figure 9-4b. This gives the metoerite an easily recognizable appearance. But these specimens are also rare and valuable because less than 2 percent of the meteoroids that strike the earth are of this variety.

The third kind of meteorite, the *irons,* is by far the easiest to find. They are composed mostly of iron but may also contain 10 to 25 percent nickel. Only 6 percent of the meteoroids that strike the earth are of this type. But because of their iron content, they stick to magnets and trigger metal detectors. For this reason, iron meteorites usually dominate the collections of museums and aficionados. Because of their abundance, you can purchase iron meteorites for only $50 per pound.*

A typical iron meteorite is shown in Figure 9-4c. This particular specimen was cut, polished, and etched with dilute nitric acid to reveal patterns of interlocking metallic crystals. These patterns, called Widmanstätten figures, are proof of the meteorite's authenticity. It is impossible to counterfeit Widmanstätten figures. The long metallic crystals can grow only if a molten lump of nickel-iron is allowed to cool very slowly, over millions of years. The majority of iron meteorites display Widmanstätten figures.

---

*Typical 1978 market price.

a

Figure 9-4a   A Stony Meteorite
*This meteorite fell in Mexico and is about the size of a tennis ball.
This particular stone is a* carbonaceous chondrite *and is composed
of the most ancient material in the solar system. (Collection of
Ronald A. Oriti; Griffith Observatory.)*

Figure 9-4b   A Stony-Iron Meteorite
This meteorite, called a pallasite, *fell in Chile. It consists of a
mixture of pieces of minerals (in this case olivine) held in a matrix
of iron-nickel metal. (Collection of Ronald A. Oriti; Griffith
Observatory.)*

Figure 9-4c   An Iron Meteorite
This meteorite, called an octahedrite, *fell in Australia. When cut,
polished, and etched with acid, these meteorites display a
characteristic pattern of interlocking metallic crystals, as shown in
this specimen. (Collection of Ronald A. Oriti; Griffith
Observatory.)*

b

c

It is generally believed that most meteorites are directly related to asteroids. Countless collisions over billions of years between asteroids in the asteroid belt produce fragments that follow all sorts of orbits about the sun. Some of these orbits intersect Earth's orbit, and eventually asteroid fragments—called meteorites—are found lying on the ground.

By analyzing and comparing reflected sunlight from asteroids with reflected light from meteorite specimens in the laboratory, it is possible to get a good idea of what asteroids are made of. The reflected light from some asteroids (about 10 percent) strongly resembles reflected light from stony-iron meteorites. These stony-iron asteroids are large (100 to 200 kilometers in diameter) and are generally in the inner portions of the asteroid belt, near the orbit of Mars. A roughly equal number of asteroids reflect light much in the same way as iron meteorites. But the vast majority (about 80 percent) of asteroids appear to be made of the same material as a particular type of stony meteorite called a *carbonaceous chondrite.*

Carbonaceous chondrites have unusually high concentrations of volatile substances such as water and also contain organic molecules. This means that these meteorites have not been heated, compressed, or otherwise significantly altered since they condensed from the primordial solar nebula 4½ billion years ago. Any substantial heating, for example, would have easily driven off the meteorite's water and permanently broken the long organic molecules. Astronomers therefore believe that carbonaceous chondrites are exceedingly ancient—far older than any rocks found on the Earth or the moon. They are unaltered specimens of the primordial material of which planetesimals were made. A carbonaceous chondrite is shown in Figure 9-4a.

By examining meteorites, we can delve into the ancient history of our solar system. For example, in the late 1970s, a team of scientists from Caltech analyzed samples of a two-ton meteorite that fell in Mexico in 1969. They found unusual anomalies involving the abundances of calcium, barium, and neodymium. According to nuclear physics, the only way that these particular anomalies could have been produced is by showering atoms with a sudden flood of nuclear particles called neutrons. In nature, a deluge of neutrons ac-

Figure 9-5 A Meteorite Crater
*On very rare occasions, an exceptionally large meteorite strikes the
earth. This crater was created 22,000 years ago by a huge meteorite
that fell in Arizona. The crater is nearly a mile across, and 25 tons
of iron meteorites have been found in the surrounding regions.
(From* Geology Illustrated *by George S. Shelton. W. H. Freeman and
Company. Copyright © 1966.)*

companies a supernova explosion. It is logical to conclude, therefore,
that a supernova explosion was responsible for the creation of our
solar system. Shock waves from the violent, cataclysmic death of a
nearby massive star were responsible for a slight compression of the
sparse interstellar medium in our neighborhood. From that point on,
gravity could lead the way to the birth of the sun and planets.

   Although most meteorites come from the asteroid belt, most
meteors do not. A *meteor*—sometimes called a "shooting star"—is
simply the streak of light produced by a small meteoroid burning up
in the earth's atmosphere. Most meteorites come from dense
meteoroids that ultimately originated in the asteroid belt. Only these

189

dense meteoroids can survive a high-speed descent through the atmosphere without being completely vaporized. In contrast, most meteors are evidently produced by fluffy, low-density meteoroids that come from comets.

A surprising amount of meteoritic material is swept up by the earth as we orbit the sun. In fact, about 3,000 tons of interplanetary dust and rocks fall on the earth *each day.* Most of this material is in the form of dust grains (called micrometeorites) that are too tiny and lightweight to produce meteors. This dust simply drifts down through the atmosphere and settles on the ground. Samples are easily collected at unpolluted locations, such as the arctic and antarctic regions. You can even obtain some of this interplanetary dust (along with unrelated junk) by washing down the roof of your home and filtering the runoff at a downspout.

Although most of the material that strikes the earth is in the form of dust, we occasionally pass through a swarm of larger particles. The result of a *meteor shower.* About a dozen meteor showers occur each year. During a shower, the meteors generally seem to come from a particular location in the sky. The shower is therefore named after the constellation nearest that part of the sky. Some of the major showers are listed on the facing page, along with their dates of maximum display.

On a clear moonless night during the peak of a shower, it is usually possible to see 60 meteors per hour. But on rare occasions, the earth passes through an exceptionally dense swarm of meteoroids. For example, during the early morning hours of November 17, 1966, more than 2,000 meteors per minute were sighted over the western United States.

The best time to observe meteors is during the early morning hours. During that time, you are located on the "leading" side of the earth, as shown in Figure 9-7. The situation is analogous to driving down a freeway during a rainstorm. Far more raindrops pelt the front windshield (the "leading" side of the car) than the back window. Meteors seen during the evening hours must come from meteoroids moving fast enough to overtake the earth.

There is a substantial amount of dust scattered around our solar system. The total mass of this fine-grained material has been

Name of shower	Date of maximum display
Lyrids	April 22
Perseids	August 12
Orionids	October 21
Leonids	November 17
Geminids	December 13

estimated at about 25 trillion tons. That is about the same as the mass of a small-sized asteroid. But because of the solar wind, the action of sunlight, and the planets revolving about their orbits, much of this interplanetary dust is constantly being swept up or blown out of the solar system. In fact, about 8 tons of dust are removed from interplanetary space *each second.* Consequently, interplanetary dust must be replenished at the rate of 8 tons per second, or else all the dust would have long ago disappeared. Some of this replenishment comes from asteroids that are being ground up and pulverized by collisions. But a substantial contribution comes from comets.

Comets are among the most beautiful objects that can be seen in the sky. More than a dozen comets are discovered each year, although most remain so faint that they are never easily visible to the naked eye. But occasionally, one of these ghostly visitors passes near enough to the sun and Earth to produce a spectacular sight. For several nights — or sometimes weeks — a beautiful flowing cometary tail is seen arching across the sky.

Comets come in all shapes and sizes. Some are short and stubby; others are long and thin. Nevertheless, some basic features are common to all comets. The solid portion of a comet, called the *nucleus,* is very small — typically less than 10 miles in diameter. The nucleus is composed of dust and ice — a "dirty iceberg," as first proposed by the American astronomer Fred L. Whipple. As the comet approaches the sun, the ices are vaporized and begin to fluoresce and

Figure 9-6   A Meteor
It is easy to photograph meteors by attaching your camera to a tripod and leaving the shutter open. Because of the earth's rotation, the stars will appear as arched trails, as shown in this example. (Ronald A. Oriti, Griffith Observatory.)

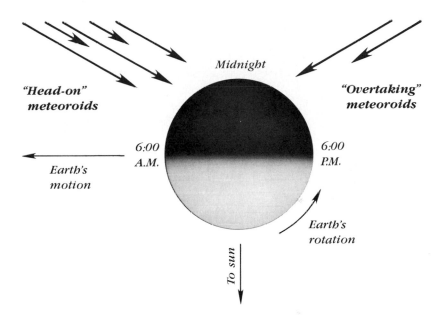

Figure 9-7   Morning vs. Evening Meteors
You always see more meteors after midnight than before. During the
early morning hours, you are on the forward-facing side of the
earth that encounters a meteoroid swarm head-on. Evening meteors
are only produced by fast-moving meteoroids that manage to catch
up with the earth.

193

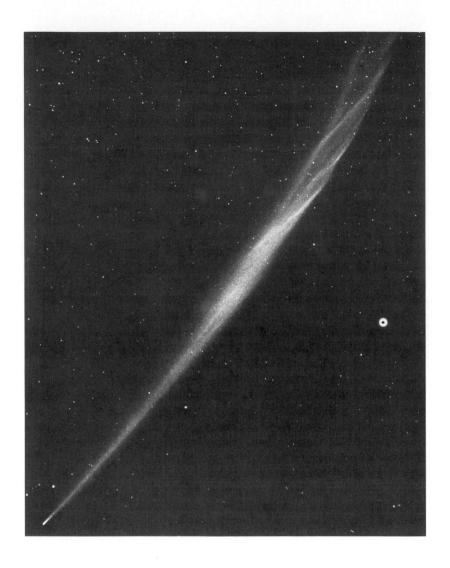

Figure 9-8   A Comet
*This comet, called Comet Ikeya-Seki after its Japanese discoverers,
was visible in October 1965. It had a very tiny head and an
extremely long tail. At the time of this photograph, the tail was
nearly 100 million miles long—about the same as the distance
between the earth and the sun. (Lick Observatory.)*

glow under the influence of the solar radiation. The nucleus is soon surrounded by a glowing ball of gases called the *coma.* Finally, as the comet plunges toward the sun, these gases are swept outward into a long, flowing *tail.* Because of the solar wind and radiation pressure, the tail always points away from the sun, as shown in Figure 9-9.

Comets approach the sun from all directions along highly elliptical orbits. Most of these orbits are so enormous that it takes 100,000 to 1 million years to complete *one* trip around the sun. But astronomers sight at least a dozen comets each year in the inner regions of the solar system (it is almost impossible to detect a comet that is farther away than Jupiter). And surely many more faint comets go completely unnoticed. This means that the total number of comets must be huge. In addition, comets spend most of their time extremely far from the sun, where they move very slowly along the most remote portions of their elongated orbits. This means that the solar system must be surrounded by a vast swarm of comets, as first suggested by the Dutch astronomer Jan H. Oort. This swarm, called the *Oort cloud,* extends out to distances of 50,000 AU (5 trillion miles, nearly a light year from the sun) and envelopes the solar system like a huge spherical halo. The Oort cloud may contain 100 billion comets.

Some comets travel along orbits so elongated that they swing past the sun only once and are flung into the depths of interstellar space—perhaps to be captured by another star. And certainly some comets do come from other planetary systems, bringing with them samples of unimagined worlds. These comets shed their dust and ice as they plummet toward the sun, thereby helping replenish our supply of interplanetary debris—debris that the earth sweeps up at the rate of 3,000 tons per day. Surely, some of the dust that settles on our towns and our homes and filters through the air we breathe comes from alien planetary systems—perhaps from worlds where creatures look toward the heavens, see what we see, and ask many of the same questions we ask.

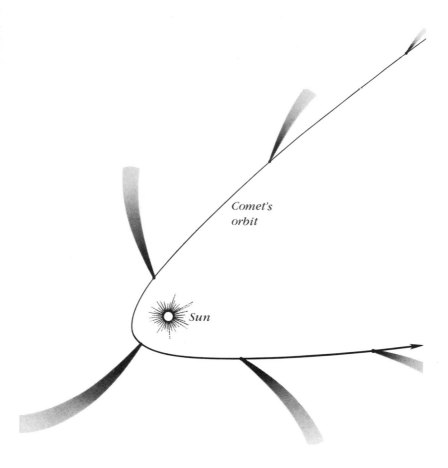

Comet's
orbit

Sun

Figure 9-9 A Comet's Orbit and Tail
Comets pass through the solar system on extremely elongated orbits.
While near the sun, ices in the comet's nucleus are vaporized
and pushed outward to form a long, flowing tail. Because of
the solar wind, the tail always points away from the sun, as shown
in this sketch.

Figure 9-10   Comet Mrkos
Some comets clearly exhibit two tails. One tail consists of ionized atoms and is pushed directly away from the sun by the solar wind. The second tail consists of dust particles and presents a hazier, arched appearance. (Hale Observatories.)

Epilogue

The building blocks for life are abundantly scattered across interstellar space. In recent years, radio astronomers have detected the weak radio signals emitted by dozens of different kinds of organic molecules. These molecules — many of which are directly linked to biological processes here on earth — are apparently so common that radio astronomers express optimism of someday being able to detect almost *any* kind of organic molecule in space. For example, at a meeting of professional astronomers in 1978, it was generally felt that it would be "just a matter of time" before amino acids are detected in interstellar clouds. Amino acids are the basic constituents of proteins that make up the DNA molecule. Several different amino acids have already been discovered inside carbonaceous chondrites.

Habitable planets are abundantly scattered around the galaxy. Although there is no direct evidence for habitable planets beyond our solar system, our understanding of the birth of stars clearly suggests that planet formation is a common process. From the scenario discussed in this book, it is entirely reasonable to suppose that planet creation accompanies the birth of all low-mass stars like the sun. Of course, many of these planets would be incapable of supporting biochemical processes. Planets too close to their parent stars have high surface temperatures that prohibit the formation of long life-giving molecules. And planets too far from their stars are so frigid that biochemical reactions simply cannot occur. Nevertheless, there must be hundreds of millions of planets that do orbit their stars at temperate distances.

199

Of course, we know of only one case where life has appeared on one of these temperate planets. And it is impossible to extrapolate or generalize on the basis of one example. But several interesting facts do stand out.

First of all, life appeared very early in the earth's history. Fossil remnants of simple one-celled organisms (resembling bacteria and blue-green algae) have been found in rocks 3½ billion years old. These bacteria/algae organisms are moderately complex and therefore must have evolved from more primitive life forms. But the earth is only 4½ billion years old. And it must have taken several hundred million years for the primordial earth to cool from its initial molten state and develop the first primitive oceans. Consequently, the first life forms made their appearance only a few hundred million years after our planet became habitable. On either a geological or an astronomical time scale, a few hundred million years is a very short interval. I believe that this ancient and relatively sudden appearance of life on earth strongly suggests that life develops on any planet where habitable conditions persist.

But what kind of life might develop? Science fiction writers have imagined an incredible variety of creatures. Are they all equally probable? Again, a possible answer comes from the terrestrial example.

All life on earth is based on the carbon atom. Some people have suggested that alien life forms might be based on the silicon atom. Or even the nitrogen atom. But terrestrial life is organized around the chemistry of carbon, in spite of the fact that silicon is considerably more abundant. Silicon is a primary ingredient of the earth's crust. And most of the earth's atmosphere is made of nitrogen. Yet, *no* silicon-based or nitrogen-based life forms exist on earth, even though they had 4½ billion years to develop. In many respects, carbon atoms are chemically the most versatile atoms in nature. They are ideally suited for construction of long, intricate molecules. And biologically active molecules *must* be long and intricate in order to code the tremendous amount of information necessary to create a living creature. For these reasons, the chemistry of carbon is the chemistry of life. I therefore believe that extraterrestrial creatures, like ourselves, are based on the carbon atom.

Finally, I believe that the natural process of evolution favors the development of intelligence. If two comparable creatures exist on a planet and one is dumb while the other is smart, then — all other things being equal — the more intelligent creature is better able to cope with its environment and ensure the survival of its offspring. Of course, this does not guarantee the longevity of the species. Sharks have inhabited our planet for hundreds of millions of years, whereas humanity's brief sojourn may be terminated within the next few decades. But the overall direction of microbes to humans was inevitable. From these considerations, it logically follows that an exceptionally large number of races of intelligent creatures have existed (do still exit?) around our galaxy.

It seems to me that it may be only a matter of time before a radio astronomer stumbles on a message from a race of advanced creatures. Indeed, some astronomers have begun systematically searching the skies for extraterrestrial communications. Yet, with a myopic display of stupidity, societies and organizations of professional astronomers still do not have any coherent policy or plan to follow if one of their colleagues should succeed. Many scientists believe that a recognizable communication from extraterrestrials would consist of the series of numbers 1, 2, 3, 5, 7, 11, 13, ... (the prime numbers) or something that amounts to $a^2 + b^2 = c^2$ (the Pythagorean theorem). And indeed, a series of dots and dashes containing this sort of information from some star system would be proof of intelligent creatures. After all, it is the kind of messages that we have deliberately broadcasted toward the stars to announce our existence. But surely there are other possibilities.

Suppose you went up to a chimpanzee and announced that $a^2 + b^2 = c^2$. It is inconceivable that the significance of your communication would evoke the tiniest spark of recognition. On the evolutionary scale, we are a few million years ahead of the chimpanzee. In exactly the same fashion, the simplest and most obvious statements of an advanced civilization might easily go completely unnoticed.

Instead, suppose that we detect a message from an alien civilization that is only slightly more advanced than we — say, only a few hundred thousand years beyond us. Their transmissions would be readily reocgnized as having an intelligent source. But the *content* of the message might be overwhelming! After all, just think of how a Cro-Magnon nobleman (or the Neolithic equivalent) would have responded to your explanation of the Pythagorean theorem. Although it might take him years to fully appreciate the meaning of $a^2 + b^2 = c^2$, your communications — with triangles, numbers, symbols, rulers, squares, lines and points — would portend a powerful new way of looking at the world. Although you were trying to communicate one simple theorem in plane geometry, your statements would reveal a dynamic new method of using the cerebral cortex. And indeed, judging from the cranial capacity of Cro-Magnon skulls, your slightly sub-human student certainly possessed an adequate supply of brain cells.

In exactly the same way, a communication from an alien civilization might be exactly like a message from the gods. One of *their* simplest transmissions might contain profound revelations in physics and mathematics. And, in trying to appreciate the full significance of the message, we might even learn a new way of thinking, a new way of utilizing the untapped resources of the human mind, a fundamentally new way of seeing reality. Every facet of civilization would be dramatically and permanently affected by a stream of dots and dashes from an unseen planet orbiting a dim and distant star.

For Further Reading

For a survey of modern astronomy, you might wish to consult an elementary college text. Six excellent such texts are listed below.

Exploration of the Universe, 3rd ed., George O. Abell (Holt, Rinehart and Winston, New York, 1975).

Exploring the Cosmos, 2nd ed., Louis Berman and J. C. Evans (Little, Brown, Boston, 1977).

Astronomy: The Structure of the Universe, William J. Kaufmann (Macmillan, New York, 1977).

The Universe Unfolding, Ivan R. King (W. H. Freeman and Company, San Francisco, 1976).

Contemporary Astronomy, Jay M. Pasachoff (Saunders, Philadelphia, 1977).

Astronomy: The Evolving Universe, Michael Zeilik (Harper & Row, New York, 1976).

In addition to general college texts, the following three books are recommended.

The Solar System (A *Scientific American* Book; W. H. Freeman and Company, San Francisco, 1975).

Exploration of the Solar System, William J. Kaufmann (Macmillan, New York, 1978).

The Cambridge Encyclopedia of Astronomy, Simon Mitton, ed., (Crown, New York, 1977).

During the recent years, a number of exceptional articles written for the layperson have appeared in popular journals. Some of the best are listed below, grouped according to subject.

Mercury

"The Significance of the Planet Mercury," William K. Hartmann, *Sky and Telescope,* vol. 51, no. 5, pp. 307–310 (May 1976).

"Mercury," Bruce Murray, *Scientific American,* vol. 233, no. 3, pp. 58–68 (September 1975).

Venus

"Venus," Andrew Young and Louise Young, *Scientific American,* vol. 233, no. 3, pp. 71–78 (September 1975).

Earth

"Convection Currents in the Earth's Mantle," D. P. McKenzie and Frank Richter, *Scientific American,* vol. 235, no. 5, pp. 72–89 (November 1976).

"The Collisoin Between India and Eurasia," Peter Molnar and Paul Tapponnier, *Scientific American,* vol. 236, no. 4, pp. 30–41 (April 1977).

"The Flow of Heat from the Earth's Interior," Henry N. Pollack and David S. Chapman, *Scientific American,* vol. 237, no. 2, pp. 60–76 (August 1977).

"The Steady State of the Earth's Crust, Atmosphere and Oceans," Raymond Siever, *Scientific American,* vol. 230, no. 6, pp. 72– 79 (June 1974).

"The Earth," Raymond Siever, *Scientific American,* vol. 233, no. 3, pp. 83– 90 (September 1975).

"The Subduction of the Lithosphere," M. Nafi Toksöz, *Scientific American,* vol. 233, no. 5, pp. 88– 98 (November 1975).

"The Earth's Mantle," Peter J. Wyllie, *Scientific American,* vol. 232, no. 3, pp. 50– 63 (March 1975).

Moon

"What's New on the Moon—I," Beran M. French, *Sky and Telescope,* vol. 53, no. 3, pp. 164– 169 (March 1977).

"What's New on the Moon—II," Beran M. French, *Sky and Telescope,* vol. 53, no. 4, pp. 258– 261 (April 1977).

"The Moon's Early History," William K. Hartmann, *Astronomy,* vol. 4, no. 9, pp. 6– 16 (September 1976).

"The Moon," John A. Woods, *Scientific American,* vol. 233, no. 3, pp. 93– 102 (September 1975).

Mars

"The Surface of Mars," Raymond E. Arvidson, Alan B. Binder, and Kenneth L. Jones, *Scientific American,* vol. 238, no. 3, pp. 76– 89 (March 1978).

"Volcanoes on Mars," Michael H. Carr, *Scientific American,* vol. 234, no. 1, pp. 32– 43 (January 1976).

"The Atmosphere of Mars," Clifford Grobstein, *Scientific American,* vol. 237, no. 1, pp. 34– 43 (July 1977).

"Mars," James B. Pollack, *Scientific American,* vol. 233, no. 3, pp. 107– 117 (September 1975).

"Phobos and Deimos," Joseph Veverka, *Scientific American,* vol. 236, no. 2, pp. 30– 37 (February 1977).

Jupiter

"New Studies of Jupiter," W. B. Hubbard and J. R. Jokipii, *Sky and Telescope,* vol. 50, no. 4, pp. 212–216 (October 1975).

"The Meteorology of Jupiter," Andrew P. Ingersoll, *Scientific American,* vol. 234, no. 3, pp. 46–56 (March 1976).

"The Galilean Satellites of Jupiter," Dale P. Cruikshank and David Morrison, *Scientific American,* vol. 234, no. 5, pp. 108–116 (May 1976).

"Jupiter," J. H. Wolfe, *Scientific American,* vol. 233, no. 3, pp. 119–126 (September 1975).

Outer Planets

"The Outer Planets," D. H. Hunten, *Scientific American,* vol. 233, no. 3, pp. 131–140 (September 1975).

"Measuring the Sizes of Saturn's Satellites," J. Veverka, J. Elliot and J. Goguen, *Sky and Telescope,* vol. 50, no. 6, pp. 356–359 (December 1975).

"Discovering the Rings of Uranus," James L. Elliot, Edward Dunham and Robert L. Millis, *Sky and Telescope,* vol. 53, no. 6, pp. 412–416 (June 1977).

Comets and Asteroids

"The Nature of Asteroids," Clark R. Chapman, *Scientific American,* vol. 232, no. 1, pp. 24–33 (January 1975).

"The Most Primitive Objects in the Solar System," Lawrence Grossman, *Scientific American,* vol. 232, no. 2, pp. 30–38 (February 1975).

"Smaller Bodies of the Solar System," William K. Hartmann, *Scientific American,* vol. 233, no. 3, pp. 143–159 (September 1975).

"Disintegration Phenomena in Comet West," Zdenek Sekanina, *Sky and Telescope,* vol. 51, no. 6, pp. 386–393 (June 1976).

"The Nature of Comets," Fred L. Whipple, *Scientific American,* vol. 230, no. 2, pp. 48–57 (February 1974).

Appendix:

Orbital, Physical, and Satellite Data

Table 1 Some Orbital Data

Planet name	Average distance from the Sun (AUs)	(in millions of kilometers)	(in millions of miles)	Orbital period (years)	(days)	Average orbital speed (km/sec)	(mi/hr)
Mercury	0.39	58	36	0.24	88	47.9	107,000
Venus	0.72	108	67	0.62	225	35.0	78,000
Earth	1.00	150	93	1.00	365	29.8	67,000
Mars	1.52	228	142	1.88	687	24.1	54,000
Jupiter	5.20	778	483	11.9		13.1	29,000
Saturn	9.54	1,427	887	29.5		9.6	22,000
Uranus	19.19	2,869	1,783	84.0		6.8	15,000
Neptune	30.06	4,498	2,795	164.8		5.4	12,000
Pluto	39.53	5,900	3,670	248.4		4.7	10,000

Table 2 Some Physical Data

Planet name	Diameter (kilometers)	Diameter (miles)	Mass (Earth = 1)	Density (gm/cm³)	Surface gravity (Earth = 1)
		(Earth = 1)			
Mercury	4,900	3,000	0.38	0.05	5.4	0.38
Venus	12,100	7,500	0.95	0.81	5.1	0.91
Earth	12,800	7,900	1.00	1.00	5.5	1.00
Mars	6,800	4,200	0.53	0.11	3.9	0.38
Jupiter	142,700	88,600	11.19	317.8	1.3	2.64
Saturn	120,800	75,000	9.47	95.2	0.7	1.13
Uranus	47,600	29,600	3.73	14.5	1.6	1.07
Neptune	44,400	27,600	3.49	17.2	2.2	1.41
Pluto	3,000	1,800	0.14	0.02	1.0	?

Table 3   Some Satellite Data

Name	Diameter (kilometers)	Diameter (miles)	Distance from planet's center (kilometers)	Distance from planet's center (miles)	Orbital period (days)	Year discovered
Satellite of Earth						
Moon	3,476	2,160	384,400	238,800	27.3	—
Satellites of Mars						
Phobos	25	16	9,300	5,800	0.3	1877
Deimos	13	8	23,500	14,600	1.7	1877
Satellites of Jupiter*						
Amalthea	150	90	181,000	112,000	0.5	1892
Io	3,640	2,260	442,000	262,000	1.8	1610
Europa	3,050	1,890	671,000	417,000	3.6	1610
Ganymede	5,270	3,280	1,071,000	665,000	7.2	1610
Callisto	4,900	3,050	1,884,000	1,170,000	16.7	1610
Leda	Very small		11,090,000	6,890,000	239	1974

Hestia	Very small		11,500,000	7,140,000	251	1904
Hera	Very small		11,750,000	7,300,000	260	1905
Demeter	Very small		11,800,000	7,330,000	264	1938
Adrastea	Very small		21,000,000	13,000,000	625	1951
Pan	Very small		23,000,000	14,000,000	714	1938
Poseidon	Very small		23,500,000	14,600,000	735	1908
Hades	Very small		23,700,000	14,700,000	758	1914

Satellites of Saturn

Janus	200	100	168,700	104,800	0.8	1966
Mimas	400	200	185,400	115,200	0.9	1789
Enceladus	600	400	238,200	148,000	1.4	1789
Tethys	1,000	600	294,800	183,100	1.9	1684
Dione	800	500	377,700	234,600	2.7	1684
Rhea	1,500	900	527,500	327,600	4.5	1672
Titan	5,800	3,600	1,223,000	760,000	15.9	1655
Hyperion	500	300	1,484,000	922,000	21.3	1848
Iapetus	1,300	800	3,563,000	2,213,000	79.3	1671
Phoebe	300	200	12,950,000	8,043,000	550.4	1898

Satellites of Uranus

Miranda	600	400	130,000	81,000	1.4	1948
Ariel	1,500	900	191,800	119,100	2.5	1851
Umbriel	1,000	600	267,300	166,000	4.1	1851
Titania	1,800	1,100	438,700	272,500	8.7	1787
Oberon	1,600	1,000	586,600	364,300	13.5	1787

(continued)

Table 3 (continued)

Name	Diameter (kilometers)	(miles)	Distance from planet's center (kilometers)	(miles)	Orbital period (days)	Year Discovered
Satellites of Neptune						
Triton	6,000	3,700	353,600	219,600	5.9	1846
Nereid	500	300	5,600,000	3,500,000	359.4	1949
Satellite of Pluto						
Charon	800	500	19,000	12,000	6.3	1978

*Data for the recently discovered fourteenth satellite are not yet available.

Index